Computational Physics
An Undergraduate's Guide

Darren Walker

Published by Pantaneto Press

Copyright © 2013 by D. J. Walker

Computational Physics: An Undergraduate's Guide

ISBN 978-0-9926368-0-7

All rights reserved. No part of this book may be reproduced or transmitted in any form or by any mean, electronic or mechanical, including photocopying, recording or by any information storage and retrieval system without the written permission of the author, except where permitted by law.

First published 2013

Printed and bound by CPI Group (UK) Ltd, Croydon, CR0 4YY

Table of Contents

1. Introduction

   1.1. Getting Started with Coding

   1.2. Getting to know Cygwin

       1.2.1. First Steps

       1.2.2. Bonjour Tout Le Monde

   1.3. The Rest of the Book

2. Getting Comfortable

   2.1. Computers: What You Should Know

       2.2.1. Hardware

       2.2.2. Software

       2.2.3. Number Representation and Precision

   2.2. Some Important Mathematics

       2.2.1. Taylor Series

       2.2.2. Matrices: A Brief Overview

3. Interpolation and Data fitting

   3.1. Interpolation

       3.1.1. Linear Interpolation

      3.1.2. Polynomial Interpolation

      3.1.3. Cubic Spline

  3.2. Data Fitting

      3.2.1. Regression: Illustrative Example

      3.2.2. Least Linear Squares: Matrix Form

      3.2.3. Realistic Example: Millikan's Experiment

4. Searching for Roots

  4.1. Finding Roots

      4.1.1. Bisection

      4.1.2. Newton-Raphson

      4.1.3. Secant

  4.2. Hybrid Methods

      4.2.1. Bisection-Newton-Raphson

      4.2.2. Brute Force Search

  4.3. So What's the Point of Root Searching...

      4.3.1. Infinite Square Well

      4.3.2. Finite Square Well

      4.3.3. Programming the Root Finder

5. Numerical Quadrature

  5.1. Simple Quadrature

      5.1.1. The Mid-Ordinate Rule

  5.1.2. The Trapezoidal Rule

  5.1.3. Simpson's Rule

 5.2. Advanced Quadrature

  5.2.1. Euler-McClaurin Integration

  5.2.2. Adaptive Quadrature

  5.2.3. Multi-dimensional Integration

6. Ordinary Differential Equations

 6.1. Classification of Differential Equations

  6.1.1. Types of Differential Equations

  6.1.2. Types of Solution and Initial Conditions

 6.2. Solving 1st Order ODEs

  6.2.1. The Simple Euler Method

  6.2.2. Modified and Improved Euler Methods

  6.2.3. The Runge-Kutta Method

  6.2.4. Adaptive Runge-Kutta

 6.3. Solving 2nd Order ODEs

  6.3.1. Coupled 1st Order ODEs

  6.3.2. Oscillatory Motion

  6.3.3. More Than One Dimension

7. Fourier Analysis

 8.1. The Fourier Series

8.2. Fourier Transforms

8.3. The Discrete Fourier Transform

8.4. The Fast Fourier Transform

    8.1.1. Brief History and Development

    8.1.2. Implementation and Sampling

8. Monte Carlo Methods

8.1. Monte Carlo Integration

    8.1.1. Dart Throwing

    8.1.2. General Integration using Monte Carlo

    8.1.3. Importance Sampling

8.2. Monte Carlo Simulation

    8.2.1. Random Walk

    8.2.2. Radioactive Decay

9. Partial Differential Equation

9.1. Classes, Boundary Values, and Initial Conditions

9.2. Finite Difference Method

    9.3.1. Difference Formulas

    9.3.2. Applications of Difference Formulas

9.3. Richardson Extrapolation

9.4. Numerical Methods To Solve PDEs

    9.4.1. The Heat Equation with Dirichlet Boundaries

Table of Contents

    9.4.2. The Heat Equation with Neumann Boundaries

    9.4.3. The Steady State Heat Equation

    9.4.4. The Wave Equation

  9.5. Finite Element Method

10. Advanced Numerical Quadrature

  10.1. General Quadrature

  10.2. Orthogonal Polynomials

  10.3. Gauss-Legendre

  10.4. Programming Gauss-Legendre

  10.5. Gauss-Laguerre Quadrature

11. Advanced ODE Solver and Applications

  11.1. Runge-Kutta-Fehlberg

  11.2. Phase Space

  11.3. Van der Pol Oscillator

    11.3.1. Van der Pol in Phase Space

    11.3.2. Van der Pol FFT

  11.4. The "Simple" Pendulum

    11.4.1. Finite Amplitude

    11.4.2. Utter Chaos?

  12.5. Halley's Comet

12.6. To Infinity and Beyond...

12.7. To the Infinitesimal and Below...

12. High Performance Computing

12.1. Indexing and Blocking

    12.1.1. Computer Memory

    12.1.2. Loopy Indexing

    12.1.3. Blocking

    12.1.4. Loop Unrolling

12.2. Parallel Programming

    12.2.1. Many (Hello) Worlds

    12.2.2. Vector Summation

    12.2.3. Overheads: Amdahl vs. Gustafson

12.3. More Stuff to Know...

Bibliography

Appendix

1 Introduction

"Computer Science is no more about computers than astronomy is about telescopes."

- E. W. Dijkstra

Computational Physics is sits at the juncture of arguably three of the cornerstone subjects of modern times; Physics, Mathematics, and Computer Science. Many see it as sitting between theoretical physics, where there is a focus on mathematics and rigorous proof, and experimental physics, which is based in taking observations and quantitative measurements. The computational physicist performs numerical experimentation within the confines of the computer environment, applying mathematics to both simulate and interpret complex models of physical systems. Just as the theoretician needs to master analytical mathematics, and the experimentalist requires a working knowledge of laboratory apparatus, so does the computational physicist need to know about numerical analysis and computer programming. Any of these skills require (lots of) practice in order to master but it is up to the physicist to know how to use them to interpret and, ultimately, understand the physical universe.

Now that we've got the grandiose opening statements out of the way shall be delve into something more practical...

1.1 Getting Started with Coding

You need two things to produce a computer program (besides a computer!!):

1. A text editor in which to write all your code in whatever language you choose;

2   A compiler to convert the code you have written into machine language (binary).

There are two methods by which you can write computer programs. The first method is via command line control whereby you explicitly type in commands to compile a source code file written in a text editor. The second method uses what is called an Integrated Development Environment (IDE) that is essentially a compiler and text editor wrapped up into one neat application, for example Microsoft's Visual Studio is an IDE. Typically, Unix based operating systems will favour command line control, whereas Windows will favour IDEs. I would suggest trying out different text editors and IDEs to discover what suits you best. If your university uses Unix based operating systems and you find it easier to code on those machines but do not want to splash out either on a Unix based machine (though the Raspberry Pi is reasonably priced) at home or make your Windows PC dual booting (i.e. it can run either a Unix OS or Windows OS on one machine) an alternative is Cygwin. Cygwin creates a Unix type feel on a Windows PC and it's free to download and install. Cygwin also comes with many different optional libraries and programs that are extremely useful to scientific programming, including the linear algebra package (LAPACK) library and Octave, a free alternative to MATLAB. If you can get your hands on a student version of MATLAB I recommend you use it as it is a very powerful programming tool and can be used to find quick programming solutions to problems, or as a first step towards a solution.

For a list of freely available text editors just search on the Wikipedia web site. Emacs is a popular programming text editor and is the default editor on most Unix based machines; Cygwin also contains the GNU version of Emacs. On Windows you could use Notepad however, it does not have any of the functionality of text editors specifically designed for coding. For example, programming languages have certain keywords reserved that have special meaning, for example `if', `for', and `while' to name but a few. Once written these keywords are automatically distinguished from the rest of the text in some way; different colour, different font, bolded, and so on. In Notepad all you'll get is the same black text on a white background, which is not useful for reading and debugging the code you have written. Notepad++ is a good (and free) programming text editor for Windows that supports multiple languages.

If you prefer to use IDEs there are a number available that are free to use. Some of these only support one language, for example Dev C++, whereas others support multiple languages, for example NetBeans or Code::Blocks. Microsoft do an 'Express' version of their Visual Studio IDE which is free to use.

In this book I have written the majority of the code using the Fortran programming language, using the GNU Emacs text editor and command line compiling from the Cygwin "shell" for Windows, which uses the GNU Fortran compiler 'gfortran'. I will not review the merits of the different programming languages here as the differences only really come into their own once you start to consider high performance computing, web applications, game programming, or other more specific applications. The basics of programming are sufficiently covered using just one language. That said, please be aware of different programming languages and how they may have an inherent advantage over, Fortran say, for different applications. For a challenge, or just for kicks, you could convert the programs in this book into another language.

1.2 Getting to know Cygwin

1.2.1 First Steps

The setup executable for Cygwin can be downloaded from the official web site *www.cygwin.com*. After it has downloaded run the executable and accept the defaults. If you are in the UK the mirror site you will want to use to download the required packages is *ftp.mirrorservices.org*. A list of various mirror services for different countries can be found on their web site at *www.cygwin.com/mirrors.html*; a link can be found on their home page. Once you have accepted the defaults you should be at the setup page. It contains a list of categories including items like Editors, Math, Graphics and so on, with a circular arrow symbol and the word 'Default' following each item. For now, click the items Editors, Devel, Math, and X11 so that the word following each changes to `Install', and click `Next'. It will likely inform you of dependent packages that are required and it is recommended you install these (in fact it is vital so that the packages you have chosen work correctly). Fortunately this is the default option and will be done automatically for you so just click `Next'. This will begin the install, which depending on the speed of your internet connection will take some time. I suggesting making a cup of tea and perhaps having a biscuit. Once finished make sure to check the box to create a shortcut on your desktop for quick access. You can always run the setup executable again to obtain as many packages as you should desire that come with Cygwin however be aware that the total download is rather large and will take up a significant portion of your hard disk.

Figure 1.1: Example of the Cygwin terminal

Introduction

Assuming you have downloaded and installed Cygwin double click the `Cygwin Terminal' shortcut icon on your Desktop. It should bring up a windowed command terminal that should look similar to Figure 1.1. The terminal prompt will be at the home directory of the current user and the prompt symbol is a dollar sign (\$). If you're a UNIX virgin then a directory is the same thing as a folder on Windows; I will use the directory nomenclature throughout this book. If you've followed the default settings of Cygwin this will be at the following location: C:\Cygwin\home\Name, where `Name' is your user name (the one you supplied to set up Windows on your PC). From here we want to make a new directory tree* in which to store all your programs you are going to write. I recommend you organise the directory tree to mimic the layout of this book. To do this at the command line prompt type the following:

mkdir ComputationalPhysics

This makes a new directory called ComputationalPhysics in your current directory. To get used to programming jargon 'mkdir' is referred to as the command and ComputationalPhysics as the argument. Some commands require more than one argument, some have optional arguments, and some require no arguments at all, as we shall see shortly. Note that I didn't leave a space between the words for the directory name argument. Typically, spaces have meaning when typing commands at the prompt so having directory names that are all one word avoids confusion. Next we want to make a new directory inside ComputationalPhysics called CH1, say, to represent this chapter. You could use the chapter title itself to name the directory but, as we shall see, this would make navigation around the tree much more tedious. To make the chapter 1 directory type the following at the command prompt.

cd ComputationalPhysics
mkdir CH1

---

* For example C:\Cygwin is the root of the Cygwin tree and the \home directory is a branch of that tree which can further branch into sub-directories, e.g., \Name

The first line changes the current or working directory to ComputationalPhysics and second makes a directory within this called CH1. We could now could just keep typing in repeats of the second line with modifications to the chapter number to create directories for all the chapters in this book. However, that would be tedious. Press the up-arrow on your keyboard. All the commands you type are saved for you to return to later if you so wish. This persists even if you exit from Cygwin and reopen at another time. There is a limit to the number of commands Cygwin can save but unless you want to return to a command you typed a very long time ago it shouldn't be a problem. This feature is very useful if you are testing or debugging code and are constantly having to recompile the program, more on this soon. You can now just keep pressing the up-arrow after each make directory command you enter and change the chapter number. To see the directories you've made type **ls** at the command prompt. This command lists all the directories and files contained within the current directory; at the moment we only have directories. If you've made a mistake at any point, say you typed an extra 9 for the chapter 9 directory, and would like to remove that directory then the command would be

rmdir CH99

You then need to remake the chapter 9 directory. If you're worried that you may accidentally remove a directory containing important files you shouldn't. The *rmdir* command will only remove directories that are empty and if you do try to remove a directory that isn't empty Cygwin will report an error message to that effect.

Change the current directory to CH1 and type the command *pwd*. What do you get? The command pwd stands for path to working directory and outputs your current location within the directory tree; very useful if you ever get lost. Those keen eyed readers will have spotted that the path contains forward slashes rather than the Windows backslash. This is part of that UNIX "feel" I mentioned earlier which uses forward slashes to separate drive, directory and file names. Remember to use forward slashes for navigation purposes when using Cygwin.

Introduction

The pwd command is telling us we are in the CH1 directory which has ComputationalPhysics as its parent, i.e., CH1 is a child of ComputationalPhysics. Now, go back to the parent directory. Stumped? You'd be amazed at the number of sources that are supposed to tell you useful Cygwin (UNIX) commands but fail to mention the most basic for moving about directory trees. To go up one directory, that is to return to the parent directory, all you type is

cd ..

Note the space between the cd and the dots. Now let's say we're back in the CH1 directory and we want to go to the CH2 directory, how do we do that in just one line?

cd ../CH2

Note the *forward* slash. The command tells the computer to change the working directory to CH2, which is contained within the parent directory (..) of the current directory (CH1). Obviously we knew where the chapter 2 directory was in the tree and could easily navigate to its location, hence the reason for having a logical layout and short directory names. What happens if you type cd on its own?

By now your command window must be full and the prompt is at or near the bottom of the terminal? Instead of using the scroll bar type the word *clear* and get the prompt back at the top of the window. Don't worry this has not deleted your command history just cleared your command window. You can use the scroll bar to move back up if you so wish, but why would you?! (up-arrow). The table below summarizes the Cygwin (Unix) commands for directory control and navigation.

Command	Examples of use
mkdir	**mkdir foo** creates a directory called foo in the current directory
	mkdir foo/bar creates a directory called bar in the directory foo
cd	**cd** changes the current directory to the home directory
	cd bar changes the current directory to child bar
	cd .. changes the current directory to one up the tree (parent)
rmdir	**rmdir foo/bar** removes directory bar from directory foo, if bar is empty
ls	**ls** lists directories and files in current directory
	ls foo lists the directories and files in child directory foo
pwd	**pwd** displays current location within the tree

I will leave it as an optional exercise for the reader to further subdivide the chapters into sectional directories.

In the next section we will cover writing, saving, compiling and running our first program from the command line. However, before we do that I just want to quickly mention how to ensure that your install of Cygwin and Windows communicate properly. First, we have to make Windows aware of the location of the all binary (executable) files that Cygwin and third party programs use to function. Click the Start button and click on the Control Panel option. Once there select the "System and Security" category and then click "System". On the left hand side you should see an "Advanced System Settings" option. Clicking this will bring up a "System Properties" window, select the "Advanced" tab, and click the

Introduction

"Environment Variables ..." button at the bottom. In the "Environment Variables" window you will see two boxes, one for user variables, the other for system variables. In the system variables box scroll down to the "Path" variable, select it and hit the edit button. **BE CAREFUL!!** If you accidently remove the current path value press cancel immediately as you will cause your system to work incorrectly. We want to add to the *end* of this path value the locations of the Cygwin binaries. Before typing anything ensure the path value is *not* highlighted and the cursor is at the *end* of the text. Now add the following:

;C:\Cygwin\usr\bin
;C:\Cygwin\usr\local\bin
;C:\Cygwin\usr\X11R6\bin

Note the semicolons separating the different items. Press the ok button. Secondly, we need to add a new variable to the system that will allow Cygwin and the third party software to display stuff in a terminal. To do this click "New..." in the system variables box and add a variable called

DISPLAY

with a value of

127.0.0.1:0.0

Keep pressing ok until you exit the "System Properties" window. To test this has had the desired effect first exit Cygwin if it is still running, and reload the terminal. Type startxwin at the command prompt. You should see a load of messages pop up in the Cygwin terminal and a new terminal will open (probably with a white background). Exit from this new terminal and at this point you may wish to clear your Cygwin terminal. Then type gnuplot at the command prompt. This should start the gnuplot program within the terminal, if it doesn't make sure you selected to install the Math category on the setup page. The command prompt should now read

gnuplot>

At the gnuplot prompt type

plot sin(x)

and you should get a nice plot of the sine function. If you don't then something has gone wrong somewhere. First, check for any typos in the path variable and in the new variable DISPLAY that we added. If there aren't any then check on the FAQ and User's Guide pages at www.cygwin.com.

1.2.2 Bonjour Tout Le Monde

The *Hello World* program is typically the first one that anyone learning a programming language gets to write. It gives us the basic syntax of a particular language and how to output something to the screen. In Cygwin there are a number of text editors available and most notable of these are Emacs and Vim. I personally prefer Emacs but you should try out the different editors available to see what suits; remember that there is no "best" text editor except the one you find most useful. Before we start change your current directory to CH1 if you are not already there. To invoke * Emacs, for instance, you would type at your command prompt

emacs&

The ampersand is required so that control is passed back to the command terminal after emacs is opened. You can invoke Emacs without the ampersand but will find the outcome mildly irritating. Type the following into your text editor:

```
PROGRAM HELLOWORLD
! Fortran Hello World program
IMPLICIT NONE
PRINT *, `Hello World'
```

---

* Invoke is the jargon for run or open a program. Makes it sound like you're casting a spell ... embrace the geek within!!

STOP
END PROGRAM

Note that the file is a currently generic text file and thus does not show any special formatting of a Fortran program. To get Emacs to do this we need to save the file as a Fortran file. To do this within Emacs you need to press Ctrl-x, followed by the usual Ctrl-s, which will bring up the "Save As" dialogue box. Here you can choose whatever name you like but it is best to choose a name that describes what the program does. For the *Hello World* program lets be sensible and save it as *helloWorld.f90*. The file extension '.f90' tells emacs and the compiler the code is written in free-format Fortran 90 rather than the older fixed-format of earlier versions of Fortran. You should now see that the words program, print, implicit none, stop and end have special highlighting. The code has to be complied into binary so that the computer can read and execute the instructions. In the command terminal make sure you are in the same directory as your helloWorld source file and type the following at the prompt

gfortran helloWorld.f90 -o helloWorld

All being well the compiler (in this case the GNU Fortran compiler) will have interpreted your source code contained within the helloWorld.f90 file and produced a binary or executable file called helloWorld.exe. The `-o' flag tells the compiler to name the executable as the text you specify after the flag; note that you do not need to specify the exe extension. If you do not type an output name the file gets saved as `a.exe' as default. To run this program type at the command prompt

./helloWorld

and it will print to the command terminal the text Hello World. This is slightly different to some UNIX environments where you can just type the executable file name.

Let's look at the code line by line to see what we've done. Every Fortran program has to start with the line `PROGRAM *program* name'. Note that Fortran is case *insensitive*, i.e., PROGRAM is the same as program is the same as Program and so on. I prefer to write the code in UPPERCASE while comments appear as normal mixed case to easily distinguish between the two. The next line is a comment line describing the function of the program. The compiler ignores these lines and they are there to help programmers explain their intentions. In Free Form Fortran you begin a comment line with an exclamation mark (!). The IMPLICIT NONE statement, although not required for this program, basically switches off any implicit compiler assumptions about the code you've written. For instance, if implicit none is not specified, any variable name that begins with a letter i through to n would be assigned as an integer type regardless of what you actually wanted. The next line is one that performs the task we would like. The next line prints the text string "Hello World" to the default output device, in our case the command terminal. STOP is included for completeness and tells the computer to stop executing the code. I'll discuss the stop command in more detail later on. The final line tells the compiler that the program ends here. Think of the PROGRAM/END PROGRAM as a pair of brackets; you can't have one without the other.

For an improved method for opening and saving a file we could have invoked Emacs in the following manner

emacs helloWorld.f90&

In this case Emacs would check for the existence of the named file in the current directory. If it doesn't find that file it opens a new file with the name specified. Emacs will know that it is a Fortran source file from the .f90 extension and will enable Fortran formatting from the start. If the file does exist it will open that file for editing.

There are a couple of things I'd like to mention here to make your (coding) life more easy. First, you may want to alias the `gfortran' to something shorter so that you don't have to type the long name every time you wish to compile code. To do this change your working directory to your home directory, then type

emacs .bash_profile&

This will open your bash profile file in the Emacs editor. This is a script file that is run every time you open the Cygwin terminal. At the end of the file add

alias gfc=gfortran

Avoid using spaces around the equals sign. You can now invoke the Fortran compiler using gfc rather than gfortran. To close the bash profile file within Emacs press Ctrl-x, Ctrl-c.

Second, the Emacs editor can be customised to your liking. Click the "options" button at the top of the editor and point to "Customize Emacs" and select "Custom Themes". This will bring up a list of available themes that have different background and font colours that you may find easier on the eye than the usual white background and dark text. Indeed, if you are dyslexic you will definitely find it easier to read light text on a dark background. My preference is the "deeper-blue" theme but try out different themes to find out what suits you the best. Remember to hit the "Save Theme Settings" button before going back to your Fortran file. To return to the helloWorld.f90 file press Ctrl-x, then Ctrl-f and at the bottom of the text editor you will find a prompt; type helloWorld.f90, hit return and you will be back at our hello world program, with your selected theme applied.

There are many other things you can do with your bash profile and Emacs editor that I will leave to the reader to play about with (use the internet as a guide). It reflects the UNIX philosophy that everything on your computer should be customisable and made personal.

1.3 The Rest of the Book

Each chapter describes an individual topic within the general subject area of computational physics. Where there is cross over between topics this has been explicitly referred to in the text. Throughout the remaining chapters there are frequent references to Fortran 90 files that contain example programs for your study and use. These can be found in the Appendix at the back of the book in the order that they appear in the text. For convenience, you can also find the code at the following web address:

http://compphysintro.wordpress.com/computational-physics-an-undergraduates-guide /

The chapters are arranged to provide some logical flow to the discovery of computational physics, starting out with the basic topics such as data fitting and root finding, and building to more advanced techniques, such as performing Fourier transforms and solving partial differential equations. At the end of each chapter are some exercises for the reader to do. These are designed to test you and to get you thinking like a physicist so don't be put off if you find them overly difficult at first. Use the resources available to you to find solutions, which includes fellow students, tutors, and professors, as well as that font of all knowledge Wikipedia – don't forget the library also.

In the Bibliography you will find a guide to more general reading around each of the topics discussed, including pointers to other introductory texts in computational physics.

2 Getting Comfortable

"Part of the inhumanity of the computer is that, once it is competently programmed and working smoothly, it is completely honest."

- Isaac Asimov

2.1. Computers: What You Should Know

Computers are machines that help solve complex or tedious numerical problems. In order to make the hardware perform such tasks it has to be programmed; in other words told what to do. Remember a computer program cannot think for itself and is only as clever as the programmer who wrote the code. Having an understanding of the underlying structure of a computer can help the programmer write clever code that takes advantage of that structure. For a comparison think about driving a car. You don't need to know how the car works at a component level in order to drive one. However, should you wish to improve the performance of the car, for racing, or rallying, or off-roading say, then you will have to know about the engine, the suspension, gearing, different types of tyres and fuels, streamlining the body work, and so forth. Computing is no different.

2.1.1. Hardware

The physical stuff that makes up your computer is called hardware (firmware if you're American) and consists of several components. The motherboard is the large printed circuit board that contains all the ports, plugs and electronics required to make the required components talk to each other. The central processing unit (CPU) handles the majority of tasks you ask of your computer. The speed at which the CPU handles these tasks is dependent on its clock frequency measured in Hertz. A 2 GHz single core processor can handle at most 2 billion* operations per second; operations may include additions, logic comparisons, and memory calls amongst others. Prior to 2004 clock frequencies were roughly doubling every 18 months. This followed the prediction made by Moore in the 1960s that the transistor density on silicon chips would double every 18 months. However, as the power consumed by the CPU goes up as clock frequency squared, and with global concerns over energy usage, the frequency of the CPU is now capped at under 4 GHz. Not to worry though. Performance of computers has continued to increase according to Moore's prediction through the use of multiple core machines. At the date of writing the current commercially available state-of-the-art is 16 cores, with most "standard" computers having 4 cores, though that is rapidly changing to 8 cores. Multiple cores allows for parallel operation, where by tasks can be handled simultaneously rather than having to been performed in a serial manner. For a more detailed look on parallel programming see Chapter 12 in this book.

For the CPU to be useful it must have a place to store information. Generally speaking there are four places for this information to be stored namely cache levels I and II[†], random access memory (RAM), and storage (almost always the hard disc drive (HDD)). This memory system has a hierarchical structure whereby the caches are the fastest but smallest memory levels and the HDD is the largest but slowest memory level. Level I cache typically has a size of several tens of kilobytes (if not hundreds of kilobytes in modern CPUs), and can be access at the full processor speed. It is split into two separate areas, one for data and one for instructions, both required by the CPU to function. Level II cache has a typical size of several megabytes and can also be access at the full processor speed. Level II cache acts a fast storage area for program code or variables required by that code. If the level II cache is filled by program code then the overflow is put into RAM. RAM typically has a size of several gigabytes (the computer this is being written on has 4 GB of RAM), but is accessed at a much slower speed than the caches (typically hundreds of MHz compared with GHz). If the RAM is

---

* Here we use the notion that one thousand million equals a billion
[†] Some processors, like the AMD Phenom range, have three levels of cache

filled then the CPU can store information on the HDD in a place called virtual memory. HDD today are immense coming in at relatively conservative 100 GB all the way up to 1 TB and beyond. However, the communication between CPU and virtual memory is limited by the speed at which data can be read from and written to the HDD (on the order of 1 MHz). This speed of access can be a bottle neck for programs requiring large portions of memory (think about that state-of-the-art game you wanted to run on your x year old computer but it just ran out of memory). Typically, memory considerations only come into play when you are dealing with high definition images, video, or 3D graphics. However, some numerical methods can produce matrices of extremely large size that must be dealt with efficiently in order for a computer to produce timely (and accurate) results. The rise of the graphics card, sometimes referred to as a graphical processing unit (GPU), has allowed for the development of some very sophisticated software without the need to use up CPU resources.

Other parts of a computer consist of input and output devices. Input devices are the things with which you communicate with the computer, e.g., the keyboard and mouse. Output devices are how the computer communicates back to you, usually through the monitor and printer. Other devices can be considered as 'slaves' being both controlled by the computer, and relaying data back to the computer on command; for instance a voltage-current meter can be made to perform this operation.

2.1.2 Software

How do you make all that hardware do something? Computers are controlled using programs, referred to as software. The main program that is run on your computer is the operating system or OS. Most people will have used Microsoft's Windows OS of some version in the past; the latest version at time of writing is Windows 8. Another widely available OS is UNIX which comes in various flavours. If you're an undergraduate you will almost certainly come into contact with a UNIX OS called Linux on your university's computer lab suite.

Programs are written in what are known as high level languages (e.g. Fortran, C++, etc.) and are complied into machine language or binary via another program called a compiler[*]. Note that programs such as MATLAB are interpreted languages that are designed to run effortlessly on multiplatform (i.e. different OSs) machines.

Before you attempt any programming please have a look at the following guidelines that may make your life easier:

1. Use your universities' or work's resources, which includes those sat next to you should you be in a computer lab or in your office. Failing an actual person who can communicate at least on some level, use the Internet. If you've got a complicated problem to solve it is very likely someone else has solved it already, and elegantly too (though never believe they managed it in one go without scratching their head at least once, drinking a lot of caffeine based beverages, and swearing on several occasions). Try not to treat their solution as a black box that takes your inputs and gives the desired outputs without at least trying to understand what the code is actually doing. That said there is a practical limit to anyone's knowledge and if it really makes no sense to you accept that it works and that, out there, somewhere, is someone much cleverer than you and, actually, you're ok with that.
2. Design your program first. Sit down, go old school with a pencil and paper, and write down the problem you are going to solve. What you want to get as the output and what are going to be your inputs. Draw a flow chart if it helps. Definitely write is out as pseudo code; English phrases that mimic actual code and describe the program's intended function line-by-line. This will, in the long run, save you time. Probably not straight away but practice makes perfect, allegedly. Now once this is done open your favourite text

---

[*] Begging the question if the compiler is program how did it get compiled?

editor/IDE and start tapping away, but be aware (or beware) ...

3. MAKE CERTAIN TO COMMENT ON YOUR CODE! Sorry for the capitals but this point is important. Can you remember what you were doing yesterday, a week ago, this time last month, last year? Comments not only let others know what you were intending with the code but also tell you what you were doing. Comments should be clear, concise, descriptive and written in plain English. Don't worry if you have more comments than code so long as the comments aren't waffle.
4. Make names descriptive. This includes programs, sub-routines, functions, and variables. If you have a large program spanning many hundreds of lines and/or several source code files you will be grateful to have given names that actually mean something. Additionally, many software companies will have their own naming convention for the various data types, structures, classes and so forth that can be defined and declared in a program. For instance, they may request that all pointer variable names must start with a 'p' or that functions start with an 'f'. Indeed, if you don't type ``IMPLICIT NONE'' in your Fortran program the compiler will assign any variable beginning with the letter 'i' through 'n' as an integer, regardless of what you actually wanted.
5. Do not be afraid to try something out. The worse thing that can happen it that your program crashes at runtime. Control-C, start over. Well ok maybe you crash the whole computer but that's what the power buttons for, isn't it?

Some suggest that before anything else you should check that the problem you want to solve is actually suited to the use of a computer to avoid wasting your time and computer resources. While this is a helpful tip for experienced programmers, I would argue that you only get into this habit after you have got comfortable actually writing computer code. Sometimes the simple problems allow you to explore writing novel and occasionally elegant or clever code that you may have missed tackling a more complex problem. There is no wasted time so long as you have learnt something new or, at the very least have discovered how *not* to do something.

2.1.3 Number Representation and Precision

As we are scientists we will be dealing with real numbers obtained from measurements. And being good scientists we will concerned with the precision of those numbers. During your A-level Physics course (or equivalent) your teacher will likely have banged on about significant figures, rounding off, and the difference between precision and accuracy, when taking measurements from experiments. S/He would have found it bemusing that you quoted every figure on your calculator when figuring out, say, the strength of gravity at the Earth's surface using a free fall technique. Using a simple stopwatch to determine time in free fall and distanced travelled measured with a ruler the best you could hope to achieve is around three significant figures, limited by human reactions on the stopwatch. You could increase this precision by using more precise equipment; say a computer control timing circuit and measuring the distance using a laser. The precision of your results depends upon the equipment used.

Computers can only express integer values exactly and are limited to a maximum integer value that can be expressed. Numbers in computers are stored as bits in binary format. A bit can have a logical value of 0 or 1 and strings of bits can be used to express integer numbers. A byte is a somewhat ambiguous term but generally means a string of 8 bits, and 4 bytes (i.e. 32 bits) make up what is generally referred to as one word. Here I refer to binary format as a big endian*, that is with the most significant bit written first at the left, as you would write decimal numbers. In contrast, little endian puts the most significant bit at the end on the right, which is a natural format when performing binary addition; the bits are in arithmetic order.

Take into consideration a byte or 8 bits. Each bit represents a power of two, starting at seven and ending with zero:

$$2^7 \ 2^6 \ 2^5 \ 2^4 \ 2^3 \ 2^2 \ 2^1 \ 2^0$$

---

* A term taken from Jonathan Swift's novel "Gulliver's Travels"

Getting Comfortable

So the decimal number 6 would be represented by 0000 0110 in binary format; the equation governing this is

$$\sum_{k=0}^{7}(2^k \times s)$$

where k represents the bit location and s the bit value. Note that binary is easily read in 4 bit strings that the astute reader may notice leads naturally to the hexadecimal format - honest. The maximum integer we can express with 8 bits is $2^8 - 1 = 255$. Note that $2^8 = 256$ numbers can be represented, one of which is zero hence the minus one. This data type is known as an 8 bit, unsigned integer. Note that colour images tend to be saved in this format with 8 bit unsigned integer values defining the three colour channels (RGB) leading to the statement that colour images have 16 million colours (256 x 256 x 256).

Negative integers may also be expressed using this format. To do this the first bit (most significant bit) is taken as the sign bit. This now leaves us with only 7 bits to represent a number which gives a maximum number of $2^7 - 1 = 127$. However, negative numbers can now be formed by taking the two's compliment of a positive number. To do this you first form the one's compliment by swapping the ones and zeros in the number, then add one to the result. For instance,

$$+6 = 0000\ 0110 \leftrightarrow 1111\ 1001 \leftrightarrow 1111\ 1010 = -6$$

Zero is still represented by all zeros in the bit locations (take the two's compliment of zero and you should still get zero). So what's the lowest number we can represent in this format? Take the largest number we can represent and take its two's compliment:

$$+127 = 0111\ 1111 \leftrightarrow 1000\ 0000 \leftrightarrow 1000\ 0001 = -127$$

However, note that the one's compliment of +127 is available for use and it is, by definition, one less than the two's compliment. Hence the lowest integer we can represent is -128. Note we have not lost any depth of numbers, we can still represent $2^8 = 256$ numbers; 128 (negative numbers) + 127 (positive numbers) + 1 (for the zero).

Larger numbers can be represented using larger bit stings. A 32-bit word length can represent a maximum unsigned integer of $2^{32} - 1 = 4{,}294{,}967{,}295$ or the signed integers in the range $[-2^{31}, +2^{31} - 1]$.

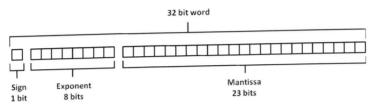

Figure 2.2: 32-bit representation of a floating point number

Computational physics would be somewhat limited if computers could only use integer numbers. We need a way of representing floating point decimal numbers. To do this we take our 32-bit word length and split it into three blocks. Figure 2.2 illustrates this representation. The first block is one bit long and represents the sign of the number, 0 and 1 representing positive and negative values, respectively. The second block, typically 8 bits long represents the exponent, and the third block, containing the remaining 23 bits is the mantissa.

The most significant bit in the *mantissa* is on the left and represents 2^{-1}. The next bit represents 2^{-2} and so forth. To calculate the floating point decimal from the 32 bit representation we use the following equation

$$x_{float} = (-1)^s \times mantissa \times 2^{exponent-bias}$$

where s is the value of the sign bit, and the *mantissa* and *exponent* are the decimal values obtained from their respective binary format blocks. The *bias* is an implicit value that is included for the following reason. The 8 bit exponent does not contain an explicit sign bit and so can only represent positive numbers up to the maximum of 255. To circumvent this drastic limitation on floating point number representation an implicit bias of 127 is included in the floating point calculation. The range of exponents hence becomes $[0 - 127 = -127, 255 - 127 = 128]$. The largest positive or negative number that can be represented is then approximately $\pm 1.7 \times 10^{38}$, and the smallest, not considering zero, is approximately $\pm 7 \times 10^{-46}$. However, do not confuse this number as the computer's precision. The computer's precision is governed by the bit length of the mantissa; the exponent just defines the range of representable numbers.

The machine precision is best described in terms of how the computer performs floating point arithmetic. Say you have the number 5 and wanted to add 10^{-7}. Both numbers can be represented by the computer in floating point notation, so far so good. To add them together the computer has to match their exponents meaning that the bits in the mantissa of the smaller number get shifted to the right. By the time the bits have been shifted to represent 10^{-7} with the same exponent as 5 they have all gone past the least significant place and have been lost, in essence making 10^{-7} equal to zero. The result of the addition would be 5.

The number 10^{-7} has not just been plucked out of thin air. The least significant bit in the mantissa has a value of $2^{-23} = 1.2 \times 10^{-7}$(2s.f.). This value represents a kind of number resolution; it is the smallest discernible difference between two numbers on a computer using a 32-bit word length. Note that it is a relative value; if you take the number $2 \times 10^6 = 2,000,000$ then the next discernible number as far as the computer is concerned is 2000000.1. The machine epsilon or precision is the unit round-off error; essentially half the number resolution. For example, numbers in the range 2000000.000 to 2000000.049 would round down to 2000000.0, whereas numbers in the range 2000000.050 to 2000000.099 would round up to 2000000.1. Any result quoted from the computer should really include this rounding error (e.g. $2,000,000 \pm 0.05$). As a consequence of the machine epsilon you should always consider whether or not the precision you are using is fit for purpose. If your calculations involve extreme differences between variable values then unit round off may lead to large errors.

Clearly the precision can be improved by adding more bits to the mantissa. This can be done by taking bits from the exponent but at the expense of the range of representable numbers. The other way of increasing the bit length of the mantissa is to double the word length from 32 to 64 bits. Both C/C++ and FORTRAN have built in types that allow the user to do this easily (double/double precision). The standard format of a double data type is an 11 bit exponent and 52 bit mantissa (plus the sign bit). What should the machine epsilon be using a double precision data type? To check your answer to that question you can write a few short lines of code to calculate the machine epsilon for both single and double precision variables. I have written the following pseudo code for this task:

Pseudo code for the calculating the machine epsilon:
Calculate the machine epsilon for both single (32 bit) and double (64 bit) precision data types.
Divide a value by 2 in a loop and test the condition that 1 plus the value is greater than 1. Break when the condition is not satisfied.

```
Program Epsilon
!!Declare the variables you are going to use
Single   eps_s = 1
Double   eps_d = 1
!!performs command while the condition is true
While( 1 + eps_s > 1 )
        eps_s = eps_s/2    !!command to execute
end While
While( 1 + eps_d > 1 )
        eps_d = eps_d/2    !!command to execute
end While
!!print results to screen
output( "single machine epsilon = ", eps_s)
output( "double machine epsilon = ", eps_d)
end Epsilon
```

Be wary that some compilers have been written to be smart, and will try to "help" when producing the binary (executable) output. For instance I wrote a C++ program using the pseudo code above and it gave a result that the single and double precision were both equal to 5.42×10^{-20}, a precision of 64 bits. This clearly is incorrect. Changing the optimisation flags I managed to recover the expected result of 5.96×10^{-8} for single precision and 1.11×10^{-16} for double precision. The incorrect result is probably due to the compiler "helpfully'" converting the variables to extended precision that has a length of 80 bits, with a 64 bit mantissa. In any case, the initial result was clearly incorrect, and that brings me to an important point. Don't just blindly accept what the computer outputs. If the answer looks wrong then it most probably is wrong. Just like when *that* guy in your (A-level) physics class stated boldly that the strength of Earth's gravity is two orders of magnitude larger than it actually is because that's what his calculator outputted, only to later realise he'd been using centimetres rather than metres.

2.2. Some Important Mathematics

Physics describes the universe from tiniest sub-atomic particle to the shape of the universe itself. The language of physics is mathematics. However do not confuse the two; Physics is not the study of mathematics (and vice versa) but uses maths as a tool to describe and interpret the observation that we make of the universe. Nowhere is this more true than when dealing with computers that are, at the most basic level, efficient number crunching machines.

In this section we will briefly review some fundamental mathematical concepts that are vital to any computational scientist performing numerical analysis. It is hoped you've have see some of these ideas before at least to some basic level.

2.2.1 Taylor Series

Brook Taylor was an English mathematician born in 1685 who devised an extremely useful way of approximating a function; the Taylor series expansion. This series expansion is arguably one of the most useful in mathematics and certainly within numerical analysis, and will play a major role in much of the subject matter contained in this book.

The Taylor series is a mathematical technique for expressing a (potentially) complicated function in the form of a polynomial. The polynomial will have a similar value to the approximated function at least in some small neighbourhood of a particular point. More precisely, a Taylor series is an infinite sum of power terms that represent a function at a single point. The summation terms are calculated from the values of the function's derivatives at that point. Mathematically we write

$$f(x) = f(a) + (x-a)f'(a) + \frac{(x-a)^2}{2!}f''(a) + \cdots$$
$$+ \frac{(x-a)^{n-1}}{(n-1)!}f^{(n-1)}(a) + \frac{(x-a)^n}{n!}f^n(\xi) \quad (2.1)$$

where a is some point on x and we have used the notation that

$$f'(x) = \frac{df}{dx}. \quad (2.2)$$

The last term in Equation (2.1) is the remainder or the error in the approximation where $a \leq \xi \leq x$.

Usually functions are approximated by using a finite number of terms of its Taylor series. Any finite number of initial terms of the Taylor series of a function is called a Taylor polynomial, the order of the polynomial governed by the highest power left in the approximation. For instance a first ordered Taylor polynomial has the form

$$f(x) \approx f(a) + (x-a)f'(a). \quad (2.3)$$

Note the use of the approximately equals to sign as we have not included the remainder term here.

You can think of the first ordered Taylor polynomial approximation as attempting to match the local neighbourhood of the function at the point $x = a$, using the function's slope at that point. The second order polynomial, then, includes the curvature of the function at the point of interest. As more terms are added, higher ordered derivatives become utilised leading to a more accurate approximation of the function around the point of interest. Typically, the approximation is only usefully accurate over a *closed interval* about the point. A function that is *equal* to its Taylor series in an *open interval* is known as an analytic function. For instant a straight line function with some non-zero gradient would be exactly given by Equation (2.3) and is thus analytic.

The upper bound to the error in a Taylor polynomial can be estimated by analysing the next term in the series from where we truncated the approximation. For example, take the Taylor series for the sine function taken about zero and truncated so that it is a seventh ordered polynomial approximation

$$\sin(x) \approx x - \frac{x^3}{3!} + \frac{x^5}{5!} - \frac{x^7}{7!}. \qquad (2.4)$$

The upper bound to the error is then calculated by the next term in the series thus

$$\varepsilon = \pm \frac{x^9}{9!}. \qquad (2.5)$$

We call this an upper bound as we have ignored any higher ordered terms, which tend to improve the accuracy of the approximation, i.e. reduce the error.

The sine function and its seventh ordered Taylor polynomial are plotted in Figure 2.1. Here we can see the approximation is only reasonably accurate on the interval $[-\pi, \pi]$.

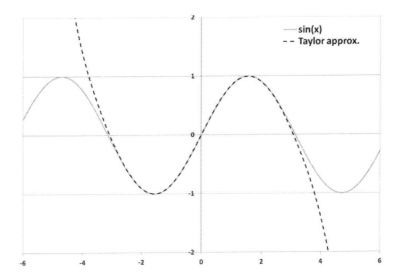

Figure 2.3: Seventh ordered Taylor polynomial approximation of the sine function.

2.2.2 Matrices: A Brief Overview

Matrices are incredibly important structures within mathematics, and thus within physics also. Here I will give a very brief over view of their form and function.

A matrix is an array of numbers. The dimensions of a matrix specify the number of rows and the number of columns the matrix has, in that order. Hence, when we say an n-by-m matrix we imply it has n rows and m columns. Vectors are essentially matrices of dimension n-by-1, for instance, a point in three-dimensional space is represented by a 3-by-1 matrix (normally referred to as a position vector). When $n = m$ we have a square matrix; these occur often in physics.

When we write algebra for matrices the notation is conventionally an uppercase letter for the entire matrix, and the corresponding lower case letter for its elements. The elements also come with numbered subscripts to denote their position within the matrix, row index first. For example the element found in the first row and the first column of matrix A would be denoted a_{11}, whereas element a_{34} is located at the third row and fourth column of A. In general element a_{ij} is found in the ith row and the jth column of the matrix A. Note that the numbering starts from one.

Diagonal elements of a matrix are identified by the fact that the row index i will equal the column index j. Sub-diagonal elements are identified by $i > j$, and conversely super-diagonal elements are identified by $i < j$. To illustrate, a general n-by-m matrix can be written as

$$A = \begin{bmatrix} a_{11} & a_{12} & \cdots & \cdots & a_{1m} \\ a_{21} & a_{22} & & & \vdots \\ \vdots & & \ddots & & \vdots \\ \vdots & & & a_{ii} & \vdots \\ \vdots & & & & \vdots \\ a_{n1} & \cdots & \cdots & \cdots & a_{nm} \end{bmatrix} \qquad (2.6)$$

where in this case $n > m$.

Matrix addition is a straight forward extension to addition with real numbers. You just add the corresponding elements between the matrices you wish to add; note that the matrices are of the same dimensions and the addition will result in a matrix also of the same dimensions. Thus for 2-by-2 matrices

$$A + B = \begin{bmatrix} a_{11} & a_{12} \\ a_{21} & a_{22} \end{bmatrix} + \begin{bmatrix} b_{11} & b_{12} \\ b_{21} & b_{22} \end{bmatrix} = \begin{bmatrix} a_{11} + b_{11} & a_{12} + b_{12} \\ a_{21} + b_{21} & a_{22} + b_{22} \end{bmatrix} = C. (2.7)$$

As this is a simple extension to addition with real numbers the properties of addition apply to matrices. In other words, we have the

Commutative property: $A + B = B + A$;

Associative property: $(A + B) + C = A + (B + C)$;

Additive Identity property: $A + ZERO = A$; and the

Distributive property: $C(A + B) = CA + CB$;

where $ZERO$ is a matrix of the same dimensions as A but every element is zero; in technical parlance this is the called the *null matrix*. Be aware that we have to be a little bit careful with the distributive property so as to maintain the proper order of the multiplication otherwise we run into problems which I will discuss next. Subtraction follows these same rules.

Matrix multiplication is somewhat more complicated than addition. If you have the n-by-m matrix A that is multiplied with the m-by-p matrix B, then the result will be the n-by-p matrix C. The elements in C are then given by the equation

$$c_{ij} = \sum_{i=1}^{n} \sum_{j=1}^{p} \sum_{k=1}^{m} a_{ik} b_{kj} \, . \tag{2.8}$$

Notice that the inner dimensions of the two matrices must match. In other words, the number of columns of matrix A must equal the number of rows of matrix B. The way Equation (2.8) has been written mimics how you will have been taught to do matrix multiplication; moving along the rows of A, and down the columns of B. However, notice that the summation limits are not dependent on each other meaning that their order could be swapped without affecting the result.

Due to the Equation (2.8), matrix multiplication is not commutative that is $AB \neq BA$. However, it is associative such that $A(BC) = (AB)C$. And we have already seen that it is distributive over matrix addition so long as you maintain strict matrix order.

When multiplying a matrix by a scalar that scalar gets *broadcast* across the entire matrix, i.e. every element gets multiplied by the scalar. As the scalar is just a number then

$$\rho(AB) = (\rho A)B = A(\rho B) = (AB)\rho \qquad (2.9)$$

where ρ is any scalar.

The trace of a square matrix is the sum of the main diagonal elements of that matrix. In equation form we write

$$tr(A) = \sum_{i=1}^{n} a_{ii} \qquad (2.10)$$

where A is a n-by-n matrix. The trace of a (square) matrix has some interesting properties not least the fact it is equivalent to the sum of the eigenvalues of the matrix A.

Eigenvalues of a matrix are related to its eigenvectors such that

$$A\underline{e} = \lambda \underline{e} \qquad (2.11)$$

Where $\underline{e}$ is an eigenvector of A and λ is its corresponding eigenvalue. Note that λ is just a number. You have probably already performed calculations using the characteristic polynomial to determine the eigenvectors and related eigenvalues of some relatively simple matrices. As a reminder the characteristic polynomial is calculated as

$$\det(A - \lambda I) = 0, \qquad (2.12)$$

which produces an n ordered polynomial in terms of the eigenvalue λ. The 'det' means calculate the determinant of the matrix contained within the brackets, which is relatively easy to do for 2-by-2, and 3-by-3 matrices but not for higher orders of matrix. I is the identity matrix that has one on its diagonal elements and zero elsewhere.

As an aside, eigenvectors and eigenvalues are important concepts with the realm of quantum physics. As an example of this the time independent Schrödinger Equation can be written in the form

$$H\psi = E\psi \tag{2.13}$$

where the matrix H represents the Hamiltonian of the system; the differential operators governing the potential and kinetic energies, ψ (pronounced psi) is the wavefunction of a quantum particle, e.g. an electron, and E is the (total) energy value of said wavefunction. In other words, ψ is the eigenvector of H, and E is the corresponding eigenvalue. If this at present makes little sense to you don't worry, just be aware that eigenvectors and eigenvalues are very important concepts in mathematics and physics.

Matrices can be transposed which means that row i is swapped with column i of the matrix. I tend to think of this as putting a double sided mirror along the diagonal of the matrix and the transpose is that which can be seen in the reflection. For instance, if we transposed our m-by-p matrix B the result would be the p-by-m matrix B^T, where the superscript T denotes the transposition. Notice that the row and column dimensions have swapped. In terms of matrix multiplication we can write

$$(AB)^T = B^T A^T. \tag{2.14}$$

If the matrices elements are complex numbers (they contain both real and imaginary terms) then we can take what is known as the Hermitian conjugate; take the complex conjugate of the elements, then transpose the matrix. In mathematical notation we write the equation

$$(AB)^\dagger = B^\dagger A^\dagger \tag{2.15}$$

where the dagger symbol ($\dagger$) denotes the Hermitian conjugation.

For square matrices there is the multiplicative identity property such that

$$AI = IA = A \tag{2.16}$$

where I is the identity matrix. When the matrix multiplication of two matrices, say X and Y, results in the identity matrix then we can say that Y must be the inverse matrix of X (or vice versa) by definition.

To compute the inverse of a matrix directly you find its matrix of cofactors and divide through by its determinant. For 2-by-2 and 3-by-3 matrices this can be done with relative ease but as the order of the matrix increases the computational effort required grows exponentially both in calculating the matrix of cofactors and finding the determinant. There are other methods for "inverting" a (square) matrix such as elimination and decomposition techniques that are much more computationally friendly. The Fortran library LAPACK has a plethora of subroutines that employ such techniques. Generally we are solving the linear set of equations

$$A\underline{x} = \underline{b} \tag{2.17}$$

where $\underline{x}$ is the vector we wish to find, $\underline{b}$ is the vector of known values, and the matrix A represents some relevant coefficients of the system we're trying to solve.

Note that if the determinant of a matrix is zero then we say the matrix is singular and non-invertible. Where say the matrix represents the coefficients of a linear system of equations this would be interpreted as the system as either having no solutions or many solutions. When the determinant is non-zero the system of equations will have exactly one unique solution.

The preceding discussions provide an (extremely) brief exposition of matrices and their properties. Hopefully for the majority of you this discussion has been quick refresher of the fundamental concepts of matrices. I would recommend you find a book dedicated to matrices and linear algebra and spend some time absorbing the material as it will definitely come in handy during your physics degree.

Exercises

1. How is the number +5 represented by a 32 bit floating point notation? Use as the equation given as a guide.

2. What happens if you change the conditional statements in the while loops to eps_s > 0 and eps_d > 0 in the machine epsilon program? Why?

3. Add to the machine epsilon code to calculate the machine precision in terms of the mantissa bit length (Hint: how many times has it divided by two?)

4. Try to write pseudo code to calculate the machine epsilon using a recursive function (a function that calls itself). Think about how to terminate the recursion.

5. Investigate the upper bound of error for the Taylor series approximation for the sine function. Is it well estimated by the next term in the series from where we truncated the series?

6. π is one of those fundamental numbers that just keeps cropping up. One way to estimate π is to analyse polygons inscribing or circumscribing the circle. For a circle of unit diameter we may formally write the expansion

$$\pi_k = \pi_\infty + \frac{c_1}{k} + \frac{c_2}{k^2} + \frac{c_3}{k^3} + \cdots$$

where k is the number of sides of the polygon, π_k is the approximation, π_∞ is the actual value of pi to be determined, and the c_i are coefficients also to be determined. Given that $\pi_8 = 3.061467$, $\pi_{16} = 3.121445$, $\pi_{32} = 3.136548$, and $\pi_{64} = 3.140331$ use the PiPolygon.f90 code provided to work out how we solve this problem for π_∞. (Tip: To compile a code using a LAPACK subroutine you need to link to that library, e.g. gfc PiPolygon.f90 −o PI −llapack.)

3 Interpolation and Data Fitting

"All programmers are playwrights and all computers are lousy actors."

– Unknown

3.1. Interpolation

3.1.1. Linear Interpolation

The principles behind interpolation and extrapolation are something every scientist should understand. Most measurements of a system, whether that is a physical experiment or theoretical calculation, will consist of pairs of discrete values; an independent variable x, which you vary, and a dependent variable y, which you measure. To extract information from these pairs of values we would, ideally, like to find an analytical function that would give us y for any arbitrary x. Often an analytical solution does not exist or is too tedious or complicated to solve. In this case how do we find a value for y that sits between measured values in x? We can either try to fit the data to some function (typically a polynomial) or interpolate the data*. The difference between the two methods is that interpolation is constrained so that the function used to approximate the data must pass through the measured data points, whereas data fitting only requires that some error function is minimised.

As the data points can be approximated by any number of functions we must have some guidelines that outline a reasonable approximation. As a general rule these guidelines usually rely on the consistency of the gradients or derivatives of the approximation and as a result may not be suitable for functions that have rapid variations, such as those with oscillatory behaviour. Sometimes, important detail about the behaviour of a function may be missed should the measurements be too sparsely spread. As a crude

---

* Should we wish to find a y beyond our measured range in x then we would *extrapolate* the data

example of this, think about measuring the displacement of a mass on a spring as a function of time. If your sample frequency (how often you take a measurement) matches the period of oscillation then your interpolated result would show the mass does not move at all, clearly an error.

Linear interpolation is probably the most intuitive method, and probably one which you use quite regularly without realising. Essentially, a straight line is assumed to approximate the function between two neighbouring data points, with the line passing through both points. Indeed, this is a fundamental concept of mathematics to find the derivative of a function; on an infinitesimally small interval any function is a straight line. Obviously on a practical level you can't make measurements that are infinitesimally distinct, the best you can possibly achieve is the precision of your measurement device.

You will probably be most familiar with the following forms of writing the equation of a straight line

$$y = mx + c \tag{3.1}$$

where m is the gradient and c is the intercept with the y axis, or

$$\frac{y - y_1}{x - x_1} = \frac{y_2 - y_1}{x_2 - x_1} \tag{3.2}$$

where the line passes through the points $(x_1, y_1), (x_2, y_2)$.

In the world of academia these equations typically take the form

$$g(x) = a_0 + a_1 x \tag{3.3}$$

where a_0 and a_1 are called the coefficients of the linear functions; they still have the same meaning as c and m respectively in the other equations. The reason for writing the coefficients as a single letter with a subscript is that it is both elegant and descriptive; the letter immediately tells you it's a coefficient rather than a variable, and the subscript tells you to which power of x the coefficient belongs. Another good reason for the subscripts is that they lend themselves quite naturally to being stored as a vector or an array in computer memory, but more on this later.

With any interpolation we are approximating the unknown function f(x) with a function g(x) with the constraint that they are equal at the measured data points which we label x_j. Thus for neighbouring data points using linear interpolation:

$$g(x_j) = f(x_j) = f_j = a_0 + a_1 x_j \tag{3.4}$$

$$g(x_{j+1}) = f(x_{j+1}) = f_{j+1} = a_0 + a_1 x_{j+1} \tag{3.5}$$

Note that they share the same coefficients as the straight line approximation is constrained to pass through both points. Solving for the coefficients (i.e. finding a in terms of f and x) the function g(x) takes the form

$$g(x) = f_j + \frac{x - x_j}{x_{j+1} - x_j}(f_{j+1} - f_j) \tag{3.6}$$

valid for the range $[x_j, x_{j+1}]$. Take a moment to verify that this equation is a straight line and equivalent to those you are familiar with. Equation 3.6 can be written in what is called symmetrical form as follows

$$g(x) = f_j \frac{x - x_{j+1}}{x_j - x_{j+1}} + f_{j+1} \frac{x - x_j}{x_{j+1} - x_j} \tag{3.7}$$

If you're currently scratching your head wondering why I've rewritten it this form it's use will become apparent in the next section discussing polynomial interpolation and Lagrange's interpolation scheme.

The code *linearInterp.f90* implements the linear interpolation scheme on the function $f(x) = sinc(x^2)$, using both the non-symmetrical and symmetrical forms of the equation*. I have done this to demonstrate that the two forms are equivalent. The output

---

* In case you haven't come across the *sinc* function before it is defined as $sinc(z) = sin(z)/z$

from this code is plotted in Figure 3.1. I have selected ten equidistant points to represent the "measured" data on the interval [0.05, 5.0]. The linear interpolation is applied to each interval pair. From the figure it can be seen that the linear interpolation does a reasonable job at approximating the function when the second and higher order derivatives are small. However, as the derivatives increase in size it becomes much less accurate. This is to be expected; the linear interpolation approximation contains no higher order terms above one and thus cannot be expected to deal with rapidly changing functions that have sizable higher order derivatives. The interpolation may be improved by taking more data points over the total range, which in essence applies the mathematical notion of the function approaching a straight line as the interval approaches zero.

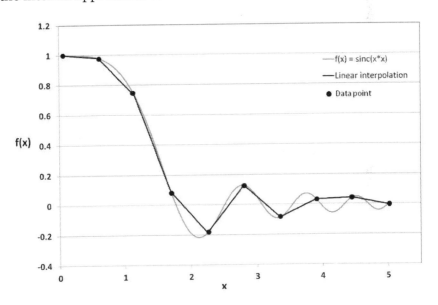

Figure 3.1: Linear interpolation of the *sinc* function using ten equidistant "measurements"

3.1.2. Polynomial Interpolation

Equation (3.3) is called a first order polynomial. By adding higher powers of x we can modify this to higher order polynomials. For instance, if the highest power of x was two it would be a second order polynomial (also called a quadratic) and so on. Higher order polynomials will be better at approximating rapidly changing functions but there is a practical limit to this, which I will get to shortly.

First, we can extend equation (3.3) so that is forms an n ordered polynomial

$$g(x) = a_0 + a_1 x + a_2 x^2 + \cdots + a_n x^n \qquad (3.8)$$

Using our interpolation constraint that the approximation must pass through the measured values gives

$$f(x_j) = f_j = g(x_j) = a_0 + a_1 x_j + a_2 x_j^2 + \cdots + a_n x_j^n \qquad (3.9)$$

This is a system of n+1 linear equations (you may know them as simultaneous equations) that we would use to solve for the coefficients. Note that to perform an n ordered interpolation you need n+1 data points. For instance, the first order (linear) interpolation requires two points; a second order interpolation requires three points, and so forth. Later in this chapter I will discuss how a linear system of equations can be solved explicitly using a LAPACK routine. For the moment we can formulate the coefficients using an alternate method.

Consider a second order interpolation for three given points (x_j, f_j) at j, j+1, and j+2:

$$f_j = a_0 + a_1 x_j + a_2 x_j^2$$
$$f_{j+1} = a_0 + a_1 x_{j+1} + a_2 x_{j+1}^2$$
$$f_{j+2} = a_0 + a_1 x_{j+2} + a_2 x_{j+2}^2 \qquad (3.10)$$

The coefficients a_0, a_1 and a_2 can be found from these equations using the methods you should have learned in an A-level mathematics course at least. Give it a go. Remember your finding the coefficients in terms of f and x. Once the coefficients are found they can be substituted into equations (3.10) and rewritten into the symmetrical form giving

$$g(x) = f_j \frac{(x-x_{j+1})(x-x_{j+2})}{(x_j-x_{j+1})(x_j-x_{j+2})} + f_{j+1} \frac{(x-x_j)(x-x_{j+2})}{(x_{j+1}-x_j)(x_{j+1}-x_{j+2})}$$
$$+ f_{j+2} \frac{(x-x_j)(x-x_{j+1})}{(x_{j+2}-x_j)(x_{j+2}-x_{j+1})} \quad (3.11)$$

If you find rearranging equations fun then feel free to have a go at obtaining this form for yourselves, but do try to get out more. If you're thinking "again with this symmetrical guff" then let me make my point I mentioned in the previous section. If you compare equation (3.7) with equation (3.11) you will hopefully see that we can generalise the symmetrical form to an n ordered polynomial interpolation scheme:

$$g(x) = f_1 \frac{(x-x_2)(x-x_3)\cdots(x-x_{n+1})}{(x_1-x_2)(x_1-x_3)\cdots(x_1-x_{n+1})}$$
$$+ f_2 \frac{(x-x_1)(x-x_3)\cdots(x-x_{n+1})}{(x_2-x_1)(x_2-x_3)\cdots(x_2-x_{n+1})} + \cdots \quad (3.12)$$
$$+ f_{n+1} \frac{(x-x_1)(x-x_2)\cdots(x-x_n)}{(x_{n+1}-x_1)(x_{n+1}-x_2)\cdots(x_{n+1}-x_n)}$$

This is the infamous Lagrange formula for polynomial interpolation. This form is somewhat cluttered and can be written more elegantly as

$$P(x) = \sum_{k=1}^{n} \lambda_k(x) f(x_k), \quad (3.13)$$

where

$$\lambda_k(x) = \frac{\prod\limits_{l=1 \neq k}^{n}(x - x_l)}{\prod\limits_{l=1 \neq k}^{n}(x_k - x_l)} \qquad (3.14)$$

The Π symbol is a capital pi and is the mathematical symbol meaning product of a sequence.

The code *lagrangeInterp.f90* implements the Lagrange polynomial interpolation for orders up to and including nine on the function $f(x) = sinc(x^2)$. Please note there is a deliberate but minor mistake within the code which one of the exercises at the end of this chapter asks you to spot and correct. Also the code is sub-optimal and can be improved, which again I ask you to do as an exercise.

Figure 3. shows the plotted output from this code for the first, second, fourth, sixth, eighth and ninth ordered polynomial interpolations. From inspection of these plots we can see that the second ordered polynomial interpolation is an improvement over the first, but still has a bad time coping as the function oscillates more rapidly. The fourth and sixth ordered interpolations are again an improvement over the second however there are two things to note. First there is a small artefact[*] within the first interval that does not follow the function at all well. Second both interpolations have only covered a fraction of the data points; in fact only the first and ninth order interpolations have covered the total number of data points. The astute amongst you will have realised this is due to the fact that an n ordered polynomial interpolation requires n+1 points. If the value of n+1 is not a factor of the total number of data points then the scheme will not be able to interpolate those points. Note the wild oscillations in the eighth and ninth ordered interpolation at the beginning and end of the interval. This is a tendency of higher order polynomial interpolation to introduce more vigorous oscillations than perhaps the data points suggest. This is the practical limit of polynomial interpolation I referred to earlier. As a rule of thumb try not to use higher than order five polynomials to do interpolation. If a

---

[*] In science a feature in the data which has been introduced by an analysis method is referred to as an artefact

greater accuracy is required you could always take more measurements, or apply another interpolation method.

To note, I have simulated measurements here by taking values from the function at equidistant points. In real measurements the data points will very likely not lie on the function that describes them due to the precision of the measuring equipment; measurements are usually plotted with their error bar. Depending on the relative size of the error this may have a significant effect on the interpolation. Additionally, we do not have to take measurements that are equidistant. I am sure that your Physics teacher told you to take more closely spaced data points about where the measured variable (y) changes rapidly with the independent variable (x).

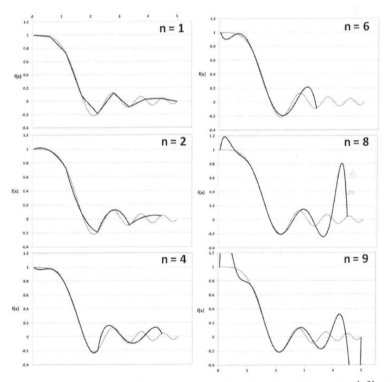

Figure 3.2: Polynomial interpolation of the function $f(x) = sinc(x^2)$ with polynomial orders of 1, 2, 4,6,8, and 9

3.1.3. Cubic Spline

One of the limiting factors of polynomial interpolation is due to the discontinuities in the derivatives at the data points (see the order 2 polynomial interpolation in figure 3.2 for a clear illustration of this issue). To overcome this issue we can employ the spline interpolation. The term spline has its origins in the ship build industry whereby thin sheets of wood threaded through discrete points (or knots) would form smooth curved shapes due to the minimisation of strain within the wood. In essence, the spline approximation not only matches the function at the measured data points but also matches the derivatives of the function at the data points.

The cubic spline is the most popular version of spline interpolation due to its (relatively) simple form and construction, and that it generally gives reasonably accurate results. The cubic part of the name comes from the order of the polynomial used to approximate the function. Cubic splines are said to have an order of four[*], which means that not only are the polynomial values matched at the data points but so are their first and second order derivatives. Given this definition the linear interpolant explored earlier in this chapter is an order two spline. What would an order one spline look like?

Cubic splines tend not to have any inherent advantage over polynomial interpolation for smooth functions or for dense sampling along the x axis. However, they are very good at interpolating sparse data points for smooth functions or when the data points vary rapidly over a region of interest, for instance, in a typical spectral measurement that contains a number of peaks and troughs. For a decent exposition of how to set up a spline approximation I recommend reading section 2.4 of T. Pang's book[ref]. It gets quite heavy on the mathematics of setting up the spline which includes generating matrices and factorising them using the Lower-Upper (LU) decomposition method. As we are physicists we like to use the

---

[*] Not to be confused with the order of the approximating polynomial

fruits of the mathematicians' labours and rather than writing our own spline approximation let's take a shortcut.

Assuming you downloaded and installed the 'Math' package form Cygwin you should have Octave installed on your machine. If not you can always rerun the setup.exe. Open the Cygwin terminal and type octave at the prompt. All being well this will open up the Octave program in your terminal. You'll know that it is working as you will see the octave prompt. Type the following at the prompt:

```
xf = [0:0.05:5];
yf = sin(xf.^2)/xf.^2;
xp=[0:0.5:5];
yp = sin(xp.^2)/xp.^2;
lin = interp1(xp, yp, xf);
spl = interp1(xp, yp, xf, "spline");
cub = interp1(xp, yp, xf, "cubic");
near = interp1(xp, yp, xf, "nearest");
plot(xf, yf, "r", xf, lin, "g", xf, spl, "b", xf, cub, ...
"c", xf, near, "m", xp, yp, "r*");
legend("function", "linear", "spline", "cubic", "nearest");
```

You should now have a neat plot of the function $y = sinc(x^2)$ with the four different types of interpolation of that function shown; the "measured" points are the red asterisks. Describing this a line at a time we have set up a line-space for x on the interval $[0,4]$ using 101 points, then calculated the function $y = sinc(x^2)$ for that line-space. Note the dot (.) before the power operator (^); this ensures the square is taken per element, rather than squaring the vector (taking it's inner product). The next line sets up a line-space of eleven points on the same interval that represent our "measured" data points, and the corresponding y is then calculated. Then we use the Octave function `interp1' to interpolate our "measured" data points using linear, cubic spline, cubic polynomial, and nearest neighbour interpolation methods. The `1' in the function name refers to the fact we are interpolating in one dimension. We then plot the results on the same figure with the legend as labelled. Note that the '...' is a continuation symbol for Octave. So all that coding in Fortran to implement the interpolation schemes, followed by importing the resulting text file into, say, Excel for plotting has been handled in ten relatively simple lines of Octave code.

This brings me back to the point I made in Chapter 1 that the problems you will encounter are likely to have already been solved, and reduced to a very elegant form. However, the idea here is not to blindly use the programs written by someone else but at least have a basic understanding of how they function. In the future you may find yourself with a problem that has yet to be tackled. In attempting a solution, you will require the skill of implementing mathematical equations in computer code, and be able to comment on the accuracy, precision, and limits of what you have written. You can only do this if you have a solid understanding of the underlying theories and equations that govern the problem and your attempted solution.

Other interpolation schemes that you may wish to investigate but are beyond the scope of this book include Rational function interpolation; B-splines; T-splines; Newton Interpolation; Neville's algorithm; and the Aitken Method. This list is not exhaustive.

3.2. Data Fitting

3.2.1. Regression: Illustrative Example

Regression is a form of data fitting that allows us to mathematically determine the line (or curve) of best fit to measured data. It is similar to interpolation in that we use measured data points to mathematically approximate a solution. However, it differs from interpolation in that instead of finding a local approximation, i.e. a function value located between two data points, we are finding the global behaviour or trend of the measurements. In that respect regression is *not* constrained to pass through the data points. In technical parlance, regression attempts to solve an over determined[*] set of simultaneous linear equations that likely have no exact solution, but will have a best fit polynomial approximation.

---

[*] More equations than unknowns

Regression methods find the coefficients of that best fit polynomial. One of the most well used regression scheme is called linear least squares where the best fit is that polynomial which minimises the sum of the squared differences between the data points and the modelled solution.

To illustrate what I mean let's have a look at a simple example. As a result of an experiment, four data points were obtained as follows: (1,2), (2,1), (3,3), (4,6) each describing an (x, y) coordinate. The experimenters want to find a line that provides the best global trend in these four data points. They initially assume that the relationship between x and y is linear and can therefore be approximated by

$$y = a_0 + a_1 x \qquad (3.15)$$

Mathematically speaking, they would like to find the numbers a_0 and a_1 that approximately solve the over determined linear system four equations in two unknowns in some "best" sense:

$$\begin{aligned} a_0 + 1a_1 &= 2; \\ a_0 + 2a_1 &= 1; \\ a_0 + 3a_1 &= 3; \\ a_0 + 4a_1 &= 6. \end{aligned} \qquad (3.16)$$

The least squares approach to solving this problem is to try to minimise the sum of the squares of the differences between the right- and left-hand sides of these equations. Putting this into algebra we are attempting to make the following function as small as possible:

$$S(a_0, a_1) = (2 - a_0 - a_1)^2 + (1 - a_0 - 2a_1)^2 + (3 - a_0 - 3a_1)^2 \\ + (6 - a_0 - 4a_1)^2. \qquad (3.17)$$

From your A-level mathematics you should remember how to find the minimum (or maximum) of a function with one independent variable; you find where the first derivative of that function is zero. For functions with multiple independent variables the method is no different only that we determine the partial derivative with respect to the independent variables separately. Apart from the variable we are taking the derivative with respect to, all other independent variables are considered constant. Applying this to the function $S(a_0, a_1)$ and after some rearrangement we obtain

$$\frac{\partial S}{\partial a_0} = 8a_0 + 20a_1 - 24 = 0$$
$$\frac{\partial S}{\partial a_1} = 20a_0 + 60a_1 - 74 = 0 \tag{3.18}$$

We know that these must give minimums and not maximums because $S(a_0, a_1) \propto a_0^2 + a_1^2$, which has no maximums. Equations (3.18) are called the normal equations and when solved give $a_0 = -0.5$ and $a_1 = 1.4$.

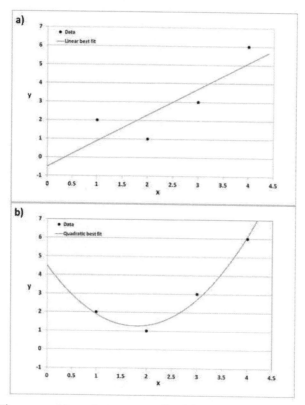

Figure 3.3: Linear least squares fit of the data using a linear model (a), and a quadratic model (b)

The line that these coefficients describe is plotted in Figure 3.3(a) along with the data points, and it is the line of best fit for a linear model. However, what if the experimenter's initial assumption about the relationship being linear is wrong? Perhaps we ought to add more terms to the approximating polynomial. Adding an extra term to our approximating polynomial gives

$$y = a_0 + a_1 x + a_2 x^2 \qquad (3.19)$$

that when processed via the method above gives us values for the coefficients of best fit of $a_0 = 4.5$, $a_1 = -3.6$, and $a_2 = 1.0$.

This curve is plotted with the data points in Figure 3.3(b). So which is the curve of best fit and thus tells the experimenters the global behaviour? Naively you may think the quadratic curve is better, it being much closer to the data points than the linear behaviour. However, upon inspection of our approximation we see that we are actually producing the Taylor series expansion for the function that passes through those specific data points. Adding more terms to the polynomial is bound to improve the accuracy of the curve passing through the points as we are providing a better approximation to the higher ordered derivatives of the function. Clearly the difficulty in interpreting the global behaviour of this simple, made up data set is due to the small number of measurements considered. Could the first data point measured at x=1 be an anomaly or an actual feature? The only way to tell in a real experiment would be to take more measurements.

3.2.2. Linear Least Squares: Matrix Form

When it comes to solving a large system of linear equations it is most convenient to write them in matrix form. The general matrix formula for a system of linear equations is

$$A\underline{x} = \underline{b} \qquad (3.20)$$

where A is a matrix of known coefficients (not to be confused with the coefficients of the approximating polynomial), $\underline{x}$ is the vector of unknown variables, and $\underline{b}$ is the vector of known right-hand side values. To illustrate this matrix form for the normal equations of the linear least squares method consider Equations (3.18). Written in matrix form they give

$$\begin{pmatrix} 8 & 20 \\ 20 & 60 \end{pmatrix} \begin{pmatrix} a_0 \\ a_1 \end{pmatrix} = \begin{pmatrix} 24 \\ 74 \end{pmatrix}.$$

We can generalise this matrix form for linear least squares to give

$$A = \begin{pmatrix} n & \sum_{i=1}^{n} x_i & \cdots & \sum_{i=1}^{n} x^m \\ \sum_{i=1}^{n} x_i & \sum_{i=1}^{n} x_i^2 & \cdots & \sum_{i=1}^{n} x^{m+1} \\ \vdots & \vdots & \ddots & \vdots \\ \sum_{i=1}^{n} x_i^m & \sum_{i=1}^{n} x_i^{m+1} & \cdots & \sum_{i=1}^{n} x_i^{2m} \end{pmatrix}, \quad (3.21)$$

where n is the total number of data points, and m is the order of the approximating polynomial.

To solve Equation (3.20), i.e. find the unknown $\underline{x}$, we have to factorise the matrix A. If you have solved a set of simultaneous equations before then you have actually factorised a matrix without realising it, probably. You will have likely used the method called Gaussian Elimination (GE) whereby you find a multiplier between two equations so as to remove a variable from their resulting addition. In matrix format this is equivalent to finding a multiplier between two rows, with the resultant addition zeroing a matrix element. With row and column exchanges the resultant matrix can be made into either a lower or upper triangular matrix[*] and we can solve the entire system from the row containing the single non-zero element on the diagonal. Note that GE is not the only factorisation method; others of note include Cholesky, LU decomposition, and QR decomposition. Rather than write our own routine to solve a system of linear equations let's take a shortcut and use a LAPACK routine to perform the factorisation. There are several LAPACK routines that

---

[*] A matrix containing only zeros above the diagonal, lower triangular, or below the diagonal, upper triangular

will perform the solve we want however I will stick with the general routine, `DGESV`, as it is adequate for our purposes and straightforward to implement. The code can be found in the program *linearLeastSquares.f90*.

Here we read in data from a text file; in this case the text file must be located in the same directory as the source code (does this have to be so?). Note that we have to allocate the size of the arrays before we can assign values to the elements. After reading in the data we set up the matrix A and the vector of right-hand-side (RHS) values $\underline{b}$. The code as written can be improved to make it automatically adjust with the order of the approximating polynomial, m, which I have left as an exercise for you to do. Once we have set up A and $\underline{b}$ all that remains to do is to solve the system of linear equations. We do this by calling the LAPACK subroutine DGESV which applies a modified version of Gaussian Elimination to solve the system. The first argument to the subroutine is the number of equations in the system, equivalent to the number of rows in matrix A. The second argument is the number of right-hand-sides, equivalent to the number of columns in $\underline{b}$. If we had multiple values for y (if measuring different properties of a physical system for the same independent variable, say) this would be greater than one. The next argument is our coefficient matrix A. The fourth argument is the leading dimension of matrix A, i.e. the number of rows in the matrix; yes this information does go in twice. IPIV is an integer array that contains information about which rows where exchanged during the factorisation; row i of the matrix was interchanged with row IPIV(i). This information is redundant for our purposes but we still have to provide the argument for the routine to work. The sixth argument is the vector of RHS, $\underline{b}$ with its leading dimension as the following argument. The last argument, INFO, is used to report back on the success (INFO= 0) or failure (INFO≠ 0) of the routine; it should return zero. The solution, i.e. the best fit polynomial coefficients, is stored in the vector $\underline{b}$, overwriting the values we pass to the function. The program prints these values to the screen before deallocating the arrays, and exiting.

To compile the code remember that we need to link the LAPACK library using the -l flag, for example:

gfc LinearLeastSquares.f90 -o LLS -llapack

We can check that the routine works by providing it with the values from our simple example and the result should be $a_0 = -0.5$ and $a_1 = 1.4$.

In its current state to increase the order of the approximating polynomial by we have to provide additional code after line 48 in the set up loop for the matrix A, namely

A(3,J) = A(3,J) + T(K)*XI(K)*XI(K)

And for another increase in order

A(4,J) = A(4,J) + T(K)*XI(K)*XI(K)*XI(K)

And so on. This gets tedious very quickly, hence the reason for exercise 4 at the end of this chapter.

3.2.3. Realistic example: Millikan's Experiment

The oil drop experiment, or more famously Millikan's Experiment, was an experiment performed by Robert Millikan and Harvey Fletcher in 1909 that provided one of the first accurate measures the elementary electric charge (the charge of the electron).

n	$q_n/10^{-19}C$	n	$q_n/10^{-19}C$
4	6.558	12	19.68
5	8.206	13	21.32
6	9.880	14	22.96
7	11.50	15	24.60
8	13.14	16	26.24
9	14.82	17	27.88
10	16.40	18	29.52
11	18.04		

Table 1: Some of Millikan's and Fletcher's oil drop data

The experiment involves balancing the gravitational force with the drag and electric forces acting on microscopic charged droplets of oil suspended between two metal electrodes. The droplet's radii can be measured, and with knowledge of the oil's density, their weight and buoyancy can be calculated. Millikan and Fletcher could use this information with a known electric field to determine the charge on oil droplets in mechanical equilibrium. By repeating the experiment for many droplets, they confirmed that the charges were all multiples of some fundamental value, and calculated it to be about 1.5924×10^{-19} C $\pm 0.01\%$. They proposed that this was the charge of a single electron.

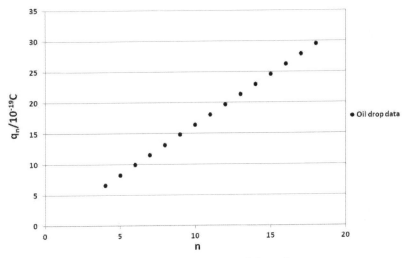

Figure 3.2: Plot of Millikan's oil drop data

Some of the data from Millikan and Fletcher's experiment are show in Table 1 and plotted in Figure 3.2. At first glance the data seem to lie perfectly on a straight line that passes through the origin. Can we show that mathematically that this is the case? Obviously the answer is yes otherwise this section would be very short!

A straight line through these data has the form

$$q_n = ne + \Delta q$$

where the fundamental charge e is the gradient of the line, and n is an integer. We can determine both e and the (estimated) error in the charge Δq from this data by applying our linear least squares program with the order of the approximating polynomial equal to one. The resulting output determines $e \approx 1.64 \times 10^{-19}$C with an estimate for the error bounds as $\Delta q \approx \pm 0.03 \times 10^{-19}$C. This is in very close agreement with the currently accepted value of $e = 1.602 \times 10^{-19}$C (3sf). To mathematically test how well the data is fitted by a straight line we can calculate what is called the residual norm. This is the square root of the sum of the squared residuals, and should be a vanishingly small number for the oil drop data presented. The other method we can employ is to increase the order of the approximating polynomial to study the relative sizes of the coefficients. Both of these methods I left as exercises for you to do.

As a cautionary note, the arguments above for least squares fitting assume that there is no error in the measurements of the independent variable x and generally speaking this assumption is valid. In fact, for the oil drop experiment the x values are necessarily integers, being multiples of the fundamental charge, and implicitly have no error. However, in some cases the error in x will comparable to the error in the measured value y. In this case, you would have to apply a total least squares approach that somehow minimises residuals in both the x and y coordinates.

Of course, sometimes you may be faced with data that is non-linear, for example data from spectral measurements or resonant phenomenon, where you will be interested in the location of a peak, and probably its width and its height. Non-linear equations are more complicated to deal with but can still be fitted in a least squares sense. The mathematics to deal with non-linear equations are beyond the scope of this book but for a decent introduction to non-linear least squares approximations see Chapter 3 of Paul L. DeVries' book "A First Course in Computational Physics".

Exercises

1. There is minor error in the Lagrange interpolation program; modify the program to correct for that error (hint: try interpolating $f(x) = cos(x^2)$)

2. Modify the Lagrange interpolation program so that each order of polynomial covers the total range of numbers.

3. Using a mathematics library see how well a cubic spline interpolation performs over polynomial interpolations for the same functions using the same data points. Choice of function and number of data points is completely free. Go nuts. Test out any other interpolation routines you may find

4. The code provided for the Linear Least Squares approximation is less than optimal.

    a. Modify the code so that the loops that set up matrix A do not have to be manually adjusted each time the user wishes to change the order (M) of the approximating polynomial. (Hint: How the matrix indices i and j relate to the power of x_i? Note that powers in Fortran are performed using a double multiply operator e.g. $x^2 = x ** 2$)

    b. Include a calculation for the residual norm

    c. Modify the code so that the order of the polynomial can be chosen by the user.

    d. Come up with a method that counts the number of data entries in a file, rather than having to explicitly store the number of entries in that file.(Tip: You can skip lines by having a read command that is blank (i.e. READ(1,*)); the next read command will start on the next line of the file).

5. Test out your modified Linear Least Squares approximation program on the Millikan data for polynomial orders greater than one. Comment on the results.

Optional: Derive the general matrix form for the normal equations. (Tip: If you're struggling start with a general polynomial of order one, work out the matrix, then expand to higher orders)

4 Searching for Roots

"However beautiful the strategy, you should occasionally look at the results."

— Sir Winston Churchill

4.1 Finding Roots

A root-finding algorithm is a numerical method, or algorithm, for finding a value x such that $f(x) = 0$, for a given function f. Such an x is called a root of the function f. This type of problem occurs often in physics and science in general typically as a starting point or intermediary process of a larger problem, though sometimes it is *the* problem.

Generally computing the root of a function cannot be done analytically and this is especially true when a function is not represented by a low order polynomial. Closed form solutions for the roots exist for polynomials up to the fourth order; I am sure you are familiar with the quadratic closed form solution[*]

$$x = \frac{-b \pm \sqrt{b^2 - 4ac}}{2a}$$

However, no such general solutions exist for order five polynomials or higher. Factorisation can be employed in finding the roots of a polynomial equation (something you will have been sick of practising for you're A-level, or equivalent, mathematics exam(s)) but tends to be viable only for well-chosen coefficients, i.e. no messy fractions to deal with. For equations that are not polynomial analytical solutions are few and far between.

Finding a root of $f(x) - g(x) = 0$ is the same as solving the equation $f(x) = g(x)$. Here, x is called the unknown in the equation. Generally speaking any equation can take the form $f(x) = 0$, so equation solving, i.e. finding x, is the same thing as computing a root

---

[*] If you've wondered how to derive this formula then just complete the square on the general form for a quadratic: $ax^2 + bx + c = 0$.

of a function. One of the first techniques you should have been taught to find the root of $f(x) = g(x)$ is to plot graphs of the two functions on the same coordinate system; the x value of where the two functions intersect is the root. Clearly this has accuracy limitations stemming from the precision of the human eye and the thickness of pencil lines. Unless you're willing to draw an absolutely massive graph this technique should be reserved for providing a rough estimate of the root, which can be passed as an initial guess to a numerical root finding algorithm. Root-finding methods, provided with an initial guess, use iteration to produce a sequence of numbers that hopefully converge towards a limit, i.e. the root you wish to find. The methods are recursive in nature, that is they compute subsequent values based on current and/or previous values of x, $f(x)$, and derivatives of $f(x)$ where appropriate.

The behaviour of root-finding algorithms is studied in numerical analysis. Algorithms perform best when they take advantage of known characteristics of the given function, and typically you find that particular algorithms perform better for particular functions. In order to evaluate the usefulness of a particular root finding method we should test its robustness in achieving reliable results, ability to find closely located roots, and rate of convergence, in that order.

4.1.1 Bisection

The bisection method is a root-finding algorithm which repeatedly bisects (halves) an interval and then selects a subinterval in which a root must lie for further processing. It is a simple and robust method, but it is also relatively slow. Because of this, it is often used to obtain a rough approximation to a solution which is then used as a starting point for more rapidly converging methods (for example, the Newton-Raphson method discussed in the next section).

In general we wish to solve $f(x) = 0$ that is defined on an interval $[a, b]$, and $f(a)$ and $f(b)$ have opposite sign. So long as $f(x)$ is continuous along this interval then the limits of the interval must contain, or bracket, at least one root. We then halve the interval size and retain the bracket that must contain a root, i.e. at the limits of the interval the value of the function has opposite sign. This step is

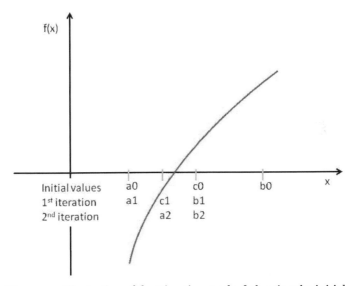

Figure 4.1: Illustration of the Bisection method showing the initial values and two subsequent iterations

repeated several times and the limits should approach the value of the root. The first two iterations of this process are illustrated in Figure 4.1. Typically, the method is repeated until some desired accuracy is achieved, or we've performed a particular number of iterations.

Explicitly, we find the midpoint between a and b

$$c = \frac{(a+b)}{2} \tag{4.1}$$

and evaluate the function at the midpoint, $f(c)$. If $f(a)$ and $f(c)$ are of opposite sign, then the method sets c as the new value for b. Else if $f(b)$ and $f(c)$ are of opposite sign the method sets c as the new value for a. In either case, the updated $f(a)$ and $f(b)$ are of opposite sign, so the method is applicable to this smaller interval. Note that if you get lucky* and $f(c) = 0$ then c is the root and the process stops.

There is a trick to determining the whether or not the function evaluations at the limits of the interval are of opposite sign without the need to assess them individually. Taking the product of two numbers with the same sign always gives a positive result. Conversely, the product of two numbers with opposite sign always gives a negative result. Hence, if the product of $f(a)$ with $f(b)$ is negative then they must be of opposite sign, if positive they must have the same sign. We can use this information in a logical condition expression of an 'IF' statement in Fortran to determine which half interval to keep and which to discard.

The program *bisection.f90* performs the bisection root finding algorithm on the function $f(x) = \cos(x) - x = 0$. This follows the programming structure in ref: DeVries pages 41-51 with some updates for the current version of the GNU Fortran compiler. A similar approach can be found in ref: Pang pages 62 - 63, which is written in Java and provides a nice comparison between the two programming languages. After compiling and running the code from this book you should find the value for the root to be $x = 0.739$ (3sf). The number of significant numbers can be adjusted by modifying the value of the parameter `TOL' (short for tolerance) within the program code. This parameter sets the minimum value of *relative* error that we will tolerate in the solution for the root.

4.1.2 Newton-Raphson

The Newton-Raphson method, named after its creators Isaac Newton and Joseph Raphson, is another method for finding successively better approximations to the roots of a function. Derivation of the method can be done by considering the geometry

---

* Not quite winning the lottery lucky but up there

of a function in the neighbourhood of the root. Consider Figure 4.2 that illustrates this point. Using the gradient of the function at an initial guess for the root we can arrive at a better approximation.

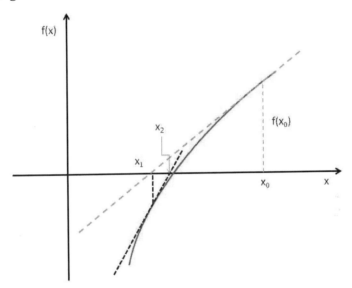

Figure 4.2: Illustration of the Newton-Raphson method show the initial value and two subsequent iterations

This is done by tracing the tangent to the function at the initial guess back to the x-axis. Repeating this method using the new value of x we arrive at an even better approximation for the root. By applying the method a number of times the approximations should converge on the root to some desired accuracy.

Using Figure 4.2 we know that

$$f'(x_0) = \frac{f(x_0)}{(x_0 - x_1)} \tag{4.2}$$

where $f'(x)$ is the gradient of the function at x. Rearranging Equation (4.2) to solve for the improved approximation to the root, x_1, we obtain

$$x_1 = x_0 - \frac{f(x_0)}{f'(x_0)} \qquad (4.3)$$

where x_0 is the initial guess. We can generalise this for the nth iteration

$$x_n = x_{n-1} - \frac{f(x_{n-1})}{f'(x_{n-1})} \qquad (4.4)$$

Note that we must be able to find the first order derivative for this method to work. This method can also be derived from the Taylor expansion of the function about the root, which I leave as an exercise for the reader.

The program *NewtRaph.f90* implements this method on the same function we looked at with the Bisection method. You can of course, change the function and its derivative to any you wish to study (so long as you can find the first derivative analytically) without the need to change anything in the subroutine. The initial guess will have to be modified so as to be close to the root of the equation you choose.

Generally, the Newton-Raphson method is good and if it converges it will converge rapidly. However, be wary that this convergence is very much dependent on the choice of the initial guess. Too far away from the root and the method will fail to converge. In some special cases the value will not converge at all; try $f(x) = x^3 - 2x + 2$ with an initial guess of one and see if you can figure out what is happening (remember CTRL-c terminates an executing program). Another issue with convergence is its reliance on the first derivative of the function in the neighbourhood of the root. What happens if the gradient of the function approaches zero at the root?

4.1.3 Secant

The Secant method uses a series of secant lines* to find better approximations of a root of a function. As with the Newton-Raphson method we can derive the recursion formula by considering the geometry of a function around the neighbourhood of the root. Figure 4.3 illustrates the derivation of the secant method.

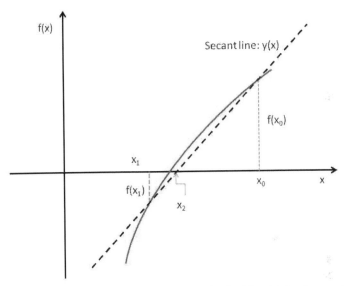

Figure 4.3: Illustration of the Secant method with initial values and one subsequent iteration

Starting with two points x_0 and x_1 that lie close to the root we draw a line through the points at $f(x_0)$ and $f(x_1)$. The equation for this straight line is given by

$$y(x) = \frac{f(x_1) - f(x_0)}{x_1 - x_0}(x - x_1) + f(x_1) \tag{4.5}$$

We wish to find the x value at which this line intersects the x-axis, in other words we find the x where $y(x) = 0$. The result is

---

* A straight line that cuts a curve in two places

$$x = x_1 - f(x_1)\frac{x_1 - x_0}{f(x_1) - f(x_0)} \qquad (4.6)$$

This value of x will be a better approximation of the root. We can then use this improved approximation, which we label x_2, with x_1 to perform the same process again to obtain an even better approximation of the root. By repeating the process iteratively we can find an approximation that lies within some desired accuracy of the actual root. We can generalise equation (4.6) for the nth iteration giving the recursion formula

$$x_n = x_{n-1} - f(x_{n-1})\frac{x_{n-1} - x_{n-2}}{f(x_{n-1}) - f(x_{n-2})}. \qquad (4.7)$$

Some readers may have noticed that equation (4.7) looks similar to the Newton-Raphson method, but with the first ordered derivative being replaced with its finite difference approximation; we look at finite difference approximations in Chapter 6. In the limit of the approximations converging on the root the second term in equation (4.7) does indeed approach the second term of the Newton-Raphson method. The Secant method should therefore be used when we do not have an analytical equation for the first derivative of the function.

Like the Newton-Raphson method the Secant method will fail to converge if the starting values are not sufficiently close to the root. It should be noted that the Secant method is more rapidly convergent than the Bisection method, but less so than for the Newton-Raphson method. Obviously, you're not just going to take my word for it, are you? Here I have not provided you with the code for the Secant method as I would like you to program it for yourself, using the Newton-Raphson code as a guide.

4.2. Hybrid Methods

4.2.1 Bisection-Newton-Raphson

From the previous section we note that the Bisection method is robust to initial guesses but slow to converge. Whereas the Newton-Raphson method will converge rapidly but only if the initial guess is relatively close to the root. Also the Newton-Raphson method may fail to converge if the gradient of the function in the neighbourhood of the root approaches zero. We would therefore like to combine the reliability of the Bisection method with the rapid convergence of the Newton-Raphson method so that for any general function we can find its roots with relative ease.

We can do this by making a hybrid method which decides whether or not to take a Newton-Raphson (NR) step or a Bisection step. As computers cannot think for themselves we as the programmers have to provide some logical criteria to determine the step to take. Crucially, if a Newton-Raphson step takes the next approximation outside of our interval then we should discard it and apply a bisection step instead; else we accept the Newton-Raphson step. To do this, let's consider our bisection interval $[a, b]$ with some best approximation to the root, r, contained within that interval. In order to accept the NR step the following inequality has to be satisfied

$$a \leq r - \frac{f(r)}{f'(r)} \leq b. \tag{4.8}$$

Now while we could use this inequality as the conditional expression in an "IF" statement the coding becomes lengthy and rather difficult to read. We can make our lives easier by rearranging the inequality in to the form

$$y \geq 0 \geq z. \tag{4.9}$$

In other words to satisfy the inequality (4.8), i.e. the NR step is accepted, the left hand expression, y, must be positive or zero, and the right hand expression, z, must be negative or zero. We can therefore apply the trick of comparing the product of y with z to zero to determine whether or not the inequality has been satisfied. To rearrange inequality (4.8) into the form of (4.9) we simply subtract r, multiply through by $-f'(r)$, and lastly subtract $f(r)$ resulting in

$$(r-a)f'(r)-f(r) \geq 0 \geq (r-b)f'(r)-f(r). \qquad (4.10)$$

If the product of the left hand side with the right hand side is negative (i.e. less than zero) or equal to zero then the NR step falls within the interval and should be accepted; else the product is positive (i.e. greater than zero) and a bisection step is applied instead. Note that a negative product will also be produced if the RHS is positive and the LHS is negative. However, this satisfies the reverse inequality of (4.10) that when tracked back to inequality (4.8) requires that $a \geq b$, which by *definition* is not the case.

The initial value for the best approximation, r, can be taken as one of the initial interval limits, it really doesn't matter. In fact, this is what we were trying to achieve; if the initial guess gives a lousy Newton-Raphson step, then the Bisection method is employed to improve the initial guess and continues to do so until the NR step falls within the interval. However, it likely that if we have reasonable guesses for the interval limits of a root, the one with the smallest function evaluation will lie closest to the root, and should be taken as the initial best guess.

The program *NRBISEC.f90* employs this hybrid method. The 'DO WHILE' loop is terminated in the usual way but contains a conditional statement to terminate the execution of the program should the desired convergence on the root not be achieved after a certain number of iterations.

If we don't have an analytic equation for the first order derivative then we should replace the Newton-Raphson method with the Secant method, so that we only have function evaluations, no derivatives. Guess what, I've left that as an exercise for you to do. Try not to stress too much; just follow the process we employed for the Newton-Raphson Bisection hybrid and you will be fine.

4.2.2 Brute Force Search

In the previous sections we have developed solid methods to accurately compute the roots of a function, so long as we know the rough locations of those roots in advance. The problem then is finding those rough locations. One straightforward technique is to graph the functions either by hand or using a plotting program and obtain those bounds by eye. This is definitely recommended when finding the roots of a function is the problem to solve. But what if the root finding is only one part of a bigger problem? It would be impractical to manually locate the rough location of roots for numerous functions in this case.

Typically we employ an exhaustive root search across a region of interest (ROI) for the function. That is, starting at the minimum value of the ROI we step the value of x by some small amount and check to see if the function has changed sign within that small step. If it has we have found the bounds of a root, if not we continue the search. This continues until the whole ROI has been covered. How then do we decide on the step size? Too small and we make our rapidly converging root finding algorithms redundant; too large and we run the risk of stepping over multiple roots (for an even number of roots this means missing them entirely; for an odd number, in essence, only one root is detected). Choosing the step size is an educated guess and is very much dependent on the function under investigation.

The subprogram *search.f90* provides an example of how to perform an exhaustive search on a function before applying the hybrid Bisection-Newton-Raphson method, say, to find the roots to some desired accuracy. Note that the specific function is the Legendre polynomial

$$P_8 = \frac{6435x^8 - 12012x^6 + 6930x^4 - 1260x^2 + 35}{128} \quad (4.11)$$

The code as it is written is very specific to this function above. It is known that P_8 has four positive roots between 0 and 1 (use a graph plotting program to confirm this if you like). Thus, an initial stab at the step size would be 0.25, but why would this not work (if you've graphed the function this should be obvious)? Try out different step sizes to see how they affect the overall performance of the program. Before leaving this section, is there any other information about the function not currently used that may help us identify the bounds for the roots of that function?

4.3. So What's The Point of Root Searching ...

We are not particularly interested in developing root searching algorithms to any great extent but we are interested in their application to physics problems we can solve with the computer. For instance, finding the roots to the Legendre polynomials is an important step in determining the evaluation points for Gaussian quadrature; very accurate methods for numerically determining the value of an integral. As a more direct example, and one we shall discuss here, finding the roots of an equation can help us calculate the energies of electrons bound in a finite square well. But first some background...

4.3.1 The Infinite Square Well

This problem is sometimes referred to as the particle in a box model. Classically the motion of the particle is governed by Newton's equations –potential fields put forces on masses causing them accelerate or change direction. At the quantum level Newton's equations are replaced by Schrödinger's such that for a particle of mass m moving through a (one dimensional) potential $V(x)$ we have

$$-\frac{\hbar^2}{2m}\frac{d^2\psi}{dx^2} = E\psi(x) - V(x)\psi(x) \qquad (4.12)$$

where $\hbar$ is Planck's constant h over $2/\pi$, E is the total energy of the system, and $\psi(x)$ is the wavefunction of the system. Note that this is the time-independent version of the Schrodinger equation; time-dependent versions also exist. Like the Newtonian equations we solve Equation (4.12) for the unknown, in this case $\psi(x)$. Although there is still some considerable debate over the nature of the wavefunction certain observable quantities do depend on its form. For instance the quantity $\psi^*(x)\psi(x)$ describes its probability function, i.e. the chance of finding the quantum particle at a particular location. More precisely the quantity $\psi^*(x)\psi(x)dx$ is the probability of finding the particle in the region x to $x + dx$.

The simplest form of the particle in a box model considers a one-dimensional system. Here, the particle may only move backwards and forwards along the x-axis with impenetrable barriers at either end. The walls of this one-dimensional box may be visualised as regions of space with an infinitely large potential energy. Conversely, the interior of the box has zero potential energy everywhere. This means that no forces act upon the particle inside the box and it can move freely in that region; remember that forces are proportional to the negative of the gradient of the potential field that causes them. If the particle touches the sides of the box it experiences an infinitely large force that pushes it back into the interior of the box; here the gradient of the potential is infinite as we have a discontinuity in the potential itself. As such the potential field is modelled by

$$V(x) = \begin{cases} 0, & 0 \leq x \leq L \\ \infty, & \text{otherwise} \end{cases} \quad (4.13)$$

where L is the length of the box and x describes the position of the particle within the box.

We now consider the wavefunction of the system both inside and outside the box. We know $\psi(x)$ must be zero outside the box as the particle is confined by the potential. Inside the box the potential is zero everywhere thus Schrodinger's equation becomes

$$\frac{d^2\psi}{dx^2} = -\frac{2m}{\hbar^2}E\psi(x). \quad (4.14)$$

The general solution of this differential equation is

$$\psi(x) = A\sin(kx) + B\cos(kx), \quad (4.15)$$

where

$$k = \sqrt{\frac{2mE}{\hbar^2}}, \qquad (4.16)$$

and A and B are constants to be determined. As it stands we currently lack the information to solve this problem. However, physical reasoning comes to our aid. We expect that the probability of finding the particle anywhere within the one dimensional space be a continuous function. As the probability of finding the particle outside the box is zero then we require the wavefunction of the particle inside the box to vanish as it approaches the walls of the box. In other words, the physics of the problem has given us the boundary conditions such that

$$\psi(0) = 0 \qquad (4.17)$$

and

$$\psi(L) = 0. \qquad (4.18)$$

Imposing these boundary conditions on the general solution at $x = 0$ we find that

$$\psi(0) = A\sin(0) + B\cos(0) = 0 \qquad (4.19)$$

which implies $B = 0$, and at $x = L$ we find that

$$\psi(L) = A\sin(kL) = 0. \qquad (4.20)$$

Equation (4.20) is satisfied either if A is zero or if $sin(kL)$ is zero. Setting $A = 0$ is rather an uninteresting case as it sets the wavefunction zero everywhere, which implies that we have no particle in our system. For $sin(kL)$ to be zero then

$$kL = n\pi \qquad (4.21)$$

where n is an integer. After substitution of Equation (4.16) and some rearrangement we find that

$$E_n = \frac{n^2 \pi^2 \hbar^2}{2mL^2}. \tag{4.22}$$

Hence, we have found a set of discrete energies that will satisfy our physical boundary conditions, and the differential equation. These are referred to eigenvalues of the system. The corresponding wavefunctions of these energies are known as the eigenfunctions. Take note that not all energies are permitted; only those that satisfy Equation (4.22) are allowed. If the particle were macroscopic (that is, not quantum) then it could take any value of (kinetic) energy it liked within the confines of the box. Figure 4.4 shows the wavefunctions and probability functions for the first four permitted energies in an infinite quantum well. Typically, we refer to E_1 as the ground state energy, and it is the lowest permitted energy the particle can attain sometimes called the zero point energy. Subsequent states we call excited states such that energy has been absorbed by the particle to jump from lower states to higher states. Note that these states are standing or stationary waves such that they are formed from two progressive waves travelling in opposite directions. These progressive waves are reflected by the infinite barrier and interact in such a way so as to produce a standing wave.

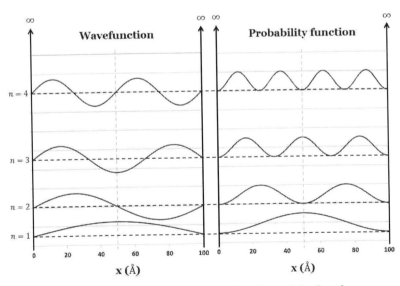

Figure 4.4: Wavefunctions and probability functions of the first four energy states of the infinite square well.

Now that we have the energies we could go back to our functions for $\psi(x)$ and find the coefficients A such that they normalise the wavefunctions, i.e.

$$\int_0^L \psi^*(x)\psi(x)dx = 1, \qquad (4.23)$$

which is the mathematical statement that the particle must be somewhere within the box. (Note that as we only consider real wavefunctions the integral function reduces to ψ^2.) In this case it is possible to show that $|A| = \sqrt{2/L}$.

The infinite square well is really only appropriate for an introduction to quantum physics. It nicely shows the discreteness of bound energy states in the well and can be solved analytically. However, as we have the computer as our disposal we could solve something a little more difficult...

4.3.2 The Finite Square Well

The finite square well is somewhat more realistic than the infinite square well. We define the potential as

$$V(x) = \begin{cases} V_0, & x < -a \\ 0, & -a \leq x \leq a \\ V_0, & x > a \end{cases} \qquad (4.24)$$

Note that in this case the well is defined symmetrically about the origin of the x-axis, rather than having a barrier at $x = 0$. We now consider the three distinct regions namely the region left of the well, the well itself, and the region right of the well. In Figure 4.5 we label these regions as I, II, and III respectively, and consider the implications of the potential field on the wavefunction in these three regions.

First, for particles with energy greater than the height of the well V_0 their wavefunctions are unbound, in other words they can move freely, and have any energy. Interestingly, as the particle moves over the well it loses potential energy, which is transformed into kinetic energy and the particle gains momentum. This manifests itself has an increase in the wavenumber of the wavefunction as the particle travels across the well (c.f. de Broglie (pronounced like Troy) momentum). This can also be seen in the differential equation. For a constant potential across x, Equation (4.12) has the form of a simple harmonic oscillator where the $E - V(x)$ term plays the role of the spring constant. As we go from region I to II the potential drops from V_0 to zero thus *increasing* the "spring constant" and the frequency of the oscillations of the particle. The opposite is true as we go from region II to III. (Strictly speaking the wave is progressive rather than stationary so we should use the time-dependent version of Equation (4.12) to govern the physics of motion, though the outcome would at least be qualitatively the same. For arguments sake, you can consider the unbound wavefunctions are the bound states of an infinitely wide quantum well.)

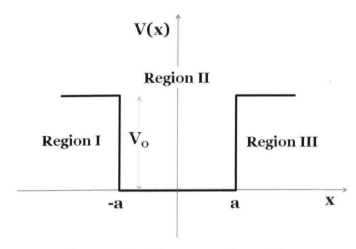

Figure 4.5: The finite square well potential

We now consider the more interesting case of particles with energy less than V_0. Starting in region I we can write the Schrödinger equation as

$$\frac{d^2\psi}{dx^2} = \frac{2m}{\hbar^2}(V_0 - E)\psi(x), \qquad (4.25)$$

which has the general solution

$$\psi_I = Ce^{\beta x} + De^{-\beta x}, \tag{4.26}$$

where

$$\beta = \sqrt{\frac{2m(V_0 - E)}{\hbar^2}}. \tag{4.27}$$

We know from experiment that the wavefunction of the particle can penetrate the finite barrier; a place where it is forbidden to go according to classical physics. If the barrier in region I had finite width then there is a probability that the particle would be found to left of region I; this is known as quantum tunnelling. We also know from experiment that the probability of finding the particle to the left of region I decreases as the width of the barrier increases, and *vanishes to zero* in the limit of the width of the *barrier* going to infinity. For the general solution to satisfy this physical observation D must be zero; remember we are in the negative half of the x-axis thus, as we go deeper into region I $e^{-\beta x}$ represents a growth function in this direction.

Moving to region II the general solution of the Schrodinger equation is the same as we found for the infinite well case restated here

$$\psi_{II} = A\sin(\alpha x) + B\cos(\alpha x), \tag{4.28}$$

where we have swapped k for α such that

$$\alpha = \sqrt{\frac{2mE}{\hbar^2}}. \tag{4.29}$$

And in region III, which identical to region II apart from the location on the x-axis, we find that

$$\psi_{III} = Fe^{-\beta x} \tag{4.30}$$

with the same reasoning for dropping the growth term. We expect that the wavefunction be continuous across x. This is again due to physical reasoning that we don't expect a sudden jump in the probability of the particles whereabouts. In addition to this we also expect that the derivative of the wavefunction to be continuous across x. In the infinite well case the discontinuity in the derivative was caused by the infinite nature of the barrier, now we have finite barriers.

To proceed we now consider the boundary conditions of the system. At $x = -a$ we obtain the following relation

$$-A\sin(\alpha a) + B\cos(\alpha a) = Ce^{-\beta a}, \qquad (4.31)$$

for the wavefunction and

$$\alpha A\cos(\alpha a) + \alpha B\sin(\alpha a) = \beta Ce^{-\beta a}, \qquad (4.32)$$

for the derivative. While at $x = a$ we find that

$$A\sin(\alpha a) + B\cos(\alpha a) = Fe^{-\beta a}, \qquad (4.32)$$

for the wavefunction and

$$\alpha A\cos(\alpha a) - \alpha B\sin(\alpha a) = -\beta Fe^{-\beta a}, \qquad (4.33)$$

for the derivative.

Taking $A = 0$ and $B \neq 0$ such that we have even parity states we find that the following must be true:

$$B\cos(\alpha a) = Ce^{-\beta a} = Fe^{-\beta a}$$

$$\Rightarrow C = F, \qquad (4.34)$$

and

$$\alpha B\sin(\alpha a) = \beta Ce^{-\beta a} = \beta B\cos(\alpha a)$$

$$\Rightarrow \alpha \tan(\alpha a) = \beta. \qquad (4.35)$$

For odd parity states where $B = 0$ and $A \neq 0$ we find similar relations:

$$-A\sin(\alpha a) = Ce^{-\beta a} = -Fe^{-\beta a}$$

$$\Rightarrow C = -F, \qquad (4.36)$$

and

$$\alpha A \cos(\alpha a) = \beta C e^{-\beta a} = -\beta A \sin(\alpha a)$$

$$\Rightarrow \alpha \cot(\alpha a) = -\beta. \qquad (4.37)$$

To find the energies and wavefunctions of the finite square well we just have to find the roots of Equations (4.35) and (4.36). And it just so happens that we have developed subroutines that can do this job...

4.3.3 Programming the Root Finder

Before launching into the code let's just remind ourselves of the nature of the problem we're trying to solve. While units like Joules, kilograms, and metres are all well and good for macroscopic objects, at the quantum level these become extremely cumbersome for quantum objects; especially when performing calculations with a computer. For example, the mass of the electron is roughly 9.1×10^{-31} kilograms and has a charge of about 1.6×10^{-19} coulombs; these are hardly numbers that will lend themselves well to precise computations. Some advocate the use of dimensionless variables such that we set a particular coefficient to unity to remove any issues of precision. For instance, we could "choose" units that set the value of $\hbar^2/2m = 1$; it doesn't matter what those units are only that the coefficient is one. However, we then have no explicit unit conversion and any results obtained are only *qualitatively* valid. It is more physically significant to use explicit unit conversions and therefore have results that are *quantifiable* also. The unit conversion will be different for different problems but the common goal is to make the coefficients have an exponent of one. Typically we can use the natural units of the problem at hand. Case in point, if we use electron masses, Angstroms, and electron-volts as our units of mass, length, and energy respectively then

$$\hbar^2 = 7.61996386 \quad m_e \text{ eV Å}^2. \tag{4.38}$$

To see how we arrived at this number let's start with the normal definition of $\hbar$ such that

$$\hbar = \frac{h}{2\pi} = \frac{6.62606957 \times 10^{-34}}{2\pi} Js, \tag{4.39}$$

and perform some dimensional analysis on the units. In SI base units the units for Planck's constant squared become

$$[J]^2[s]^2 = [kg]^2[m]^4[s]^{-2}, \tag{4.40}$$

which we can rearrange to give

$$[kg]^2[m]^4[s]^{-2} = [kg][m]^2[kg][m]^2[s]^{-2} = [kg][m]^2[J]. \tag{4.41}$$

To convert to our computer friendly units we note the following conversions:

$1 \, m_e = 9.10938291 \times 10^{-31}$ kg;

$1 \, eV = 1.60217657 \times 10^{-19}$ J;

and

$1 \, \text{Å} = 1 \times 10^{-10}$ m.

Combining these to compute $\hbar^2$ in Equation (4.38) we have

$$\hbar^2 = \frac{6.62606957^2}{4\pi^2 \times 9.10938291 \times 1.60217657} \times \frac{10^{-68}}{10^{-31} \times 10^{-19} \times 10^{-20}}$$

$$\Rightarrow \hbar^2 = 0.0761996386 \times 10^2 \quad m_e \text{ eV Å}^2. \tag{4.42}$$

By explicitly setting these unit conversions we know that when defining the well width say we do so in units of Angstroms. Or when defining the potential barrier height or computing the energy of the bound electrons we are using units of electron-volts.

Let's imagine we are attempting to find the energy and wavefunction of the lowest bound state; the ground state. From the results of the infinite square well case we would expect this state to be of even parity. Even parity states have the characteristics that $\psi \neq 0$ and $\psi' = 0$ at the middle of the well. Odd parity states have those characteristics reversed. We are therefore trying to find the energy E which satisfies the following equation

$$f(E) = \alpha \tan(\alpha a) - \beta = 0, \tag{4.43}$$

where we remind you that

$$\alpha = \sqrt{\frac{2mE}{\hbar^2}}, \tag{4.44}$$

and

$$\beta = \sqrt{\frac{2m(V_0 - E)}{\hbar^2}}. \tag{4.45}$$

Now while Equation (4.43) is perfectly acceptable as a mathematical object note that it contains properties that are abhorrent to a computer. Specifically, the tangent function contains singularities whenever $ka = n\pi$, where n is an integer, due to the cosine function being zero at these points. We can circumvent this issue by rewriting Equation (4.43) in its component terms such that

$$f(E) = \beta \cos(\alpha a) - \alpha \sin(\alpha a) = 0. \tag{4.46}$$

Here we have removed the singularities and the computer thanks us for that.

The root finding subroutines we've developed to date demand that a root be bracketed. So where do we start looking? We know that our bound wavefunctions must exist (if they exist at all) within the confines of the well. That is, they must exist only for energies between zero and the height of the potential barrier V_0. We could perform an exhaustive search on $f(E)$ but let's see if we can't do a little better by plotting the function on which we wish to perform the root search. Figure 4.6 shows the function $f(E)$ for an electron bound in a well with the parameters $a = 5$ Å, and $V_0 = 10$ eV; remember that the electron can only have energies that are equal to the roots of this function. We can clearly see three roots; the first between 0.0 eV and 0.5 eV, the second between 2.5 eV and 3.0 eV, and the third between 7.0 eV and 7.5 eV.

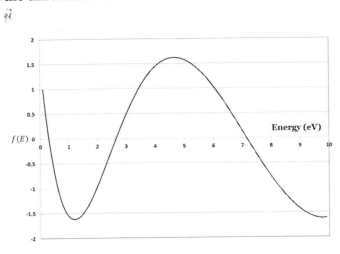

Figure 4.6: Function of energy where the roots define the energy eigenvalues for a finite square well of width **10** Å and height **10** eV

Now here's a slight rub to this plotting argument; we've plotted the function in order to see roughly where the roots lie and to avoid an exhaustive search where we would have to perform many function evaluations. However, to plot the function we have had to perform function evaluations anyway at equally spaced points and passed that data to Excel, say, for plotting. We haven't actually saved ourselves any effort and in fact have added some. In other words, we may as well perform the exhaustive search. If so desired, we could then store the function evaluations during the exhaustive search for plotting after the program has finished, providing some insight into the validity of our numerical results. As a *rule of thumb* we should set the search step length no larger than 10^{-2} of our search range to avoid skipping over roots, and no smaller than 10^{-4} of our search range to avoid an excessive number of function evaluations and subverting the intent of the root searching subroutines we've developed; if you consider ten thousand not being an excessive number! Note that the step length for an exhaustive search will very much depend upon the function being investigated. For instance, we can clearly see from Figure 4.6 that a step length of one would find brackets for all three roots.

Once we have the brackets for the roots they are passed to a root searching algorithm, say our Bisection-Secant hybrid method, for further refinement up to an accuracy specified by a user defined tolerance. We now have the energies (in eV) of the bound states of our finite square well. All that remains to do is substitute these values back into our equations for the original problem to determine their corresponding wavefunctions.

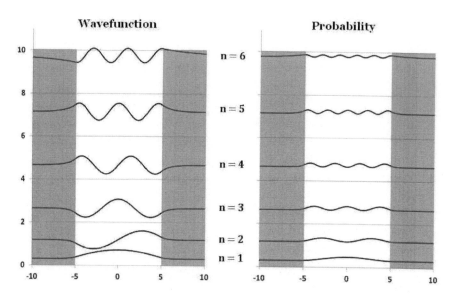

Figure 4.7: The wavefunctions and probability functions for a finite square well with the parameter defined in the text (units are Å and eV)

The file *finiteWell1.f90* contains the code to implement the discussion above for a finite square well of width 10 Å, and barrier height of 10 eV. Figure 4.7 plots the results of this code for an electron trapped in the well. Here we plot the *normalised* wavefunction on the left (see Exercise 5), and the corresponding probability function on the right; the zero baselines are aligned to their matching energy eigenvalue. This result will come in very handy as a check when we attempt to find the bound states for an arbitrary potential $V(x)$ in Chapter 11 on an advanced ordinary differential equation solver.

Exercises

1. Modify the Bisection, Newton-Raphson, and Secant method programs to determine the number of iterations required to find the root of the equation $f(x) = cos(x) - x$ to an accuracy of 8 significant figures. Comment on the dependence of conversion on the choice of initial guesses. (Tip: You may want to include a conditional exit to avoid infinite loops).

2. Modify the hybrid method programs so that it prints to screen when it's taking a bisection step and when it taking a NR/secant step. Using the various initial guess values from the previous exercise perform the same root finding computation using one of the hybrid methods we have developed. Verify that the hybrid methods are robust to initial guess and comment on how often bisection is used in comparison with the other method.

3. Modify the brute force search so that instead of attempting to find a set number of roots it only searches over a given interval of interest, reporting back the number of roots found in that interval, as well as the bracket for each.

4. The Lennard-Jones potential describes the approximate interaction between a neutral pair of atoms and has the form

Searching for Roots

$$V_{LJ} = 4\varepsilon\left[\left(\frac{\sigma}{r}\right)^{12} - \left(\frac{\sigma}{r}\right)^{6}\right]$$

where r is the distance between the atoms, and ε and σ are properties of the potential to be determined. At what value of r does the potential V_{LJ} equal zero? The size and nature of the force between the atoms is given by the magnitude and sign of the first order derivative of the potential with respect to r. At what value of r do the forces balance between the atoms, and what is the value of V_{LJ} at this point? Confirm these results using the root finding programs we have developed in this chapter. As a bonus question, what is the minimum energy required to tear the atom pair apart according to the Lennard-Jones potential?

5. The normalising coefficients for the finite square well can be calculated from

$$\int_{-\infty}^{\infty} \psi^2(x)dx = 1.$$

By performing this integral, or through research, find formulas for the coefficients A, B, and C that normalise the finite well wavefunctions. Modify the code found in *finiteWell1.f90* to reproduce the data found in Figure 4.7.

6. Study the effects of varying the well width and well height on the bound states on the bound states in the well. Do we always get at least one state, i.e. the ground state?

5. Numerical Quadrature

"To err is human--and to blame it on a computer is even more so."
— Robert Orben

Numerical integration constitutes a broad family of algorithms for calculating the numerical value of a definite integral. The term is also sometimes used to describe the numerical solution of differential equations that are described in Chapter 6 of this book. This chapter focuses on the calculation of definite integrals. The term numerical quadrature (often abbreviated to just quadrature) is more or less a synonym for numerical integration, especially as applied to one-dimensional integrals.

The basic problem considered by numerical integration is to compute an approximate solution to a definite integral:

$$\int_a^b f(x)dx. \qquad (5.1)$$

If $f(x)$ is a smooth, well-behaved function and the limits of the integration are bounded, there are several methods of approximating the integral using numerical integration to a desired precision.

The need for numerical integration in physics will become ever more apparent during your degree course and beyond. As physicists, you will likely encounter many nasty looking integrations that will either have no analytical solution or have such complicated solutions it will make you wish you'd done that degree in Media Studies[*]. Normally, we will only want the value of the integration within a bounded interval. Numerical integration will be your friend in this case.

The first two numerical integration schemes we discuss next should be very familiar to you and provide intuitive and illustrative examples of what all numerical integrations schemes are doing regardless of their complexity.

---

[*] Whatever they are?!

5.1. Simple Quadrature

5.1.1 The Mid-Ordinate Rule

The mid-ordinate rule computes an approximation to a definite integral, made by finding the area of a collection of rectangles whose heights are determined by the values of the function at certain discrete, evaluation points along the interval.

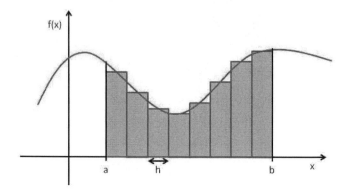

Figure 5.1: Illustration of the mid-ordinate rule

Figure 5.1 illustrates the mid-ordinate method. Specifically, the interval $[a, b]$ over which the function is to be integrated is divided into N equal subintervals of length $h = (b - a)/N$. The height of the rectangle is then determined to be the value of the function found at the mid-point between each subinterval hence the name. The approximation to the integral is then calculated by adding up the areas (base multiplied by height) of the N rectangles, giving the formula:

$$\int_a^b f(x)dx \approx h \sum_{n=0}^{N-1} f(x_n) \qquad (5.2)$$

where $x_n = a + (n + 1/2)h$. As N gets larger, this approximation gets more accurate. As N approaches infinity, h becomes infinitesimally small (approaches dx) and we have the definition of an integral. Why is this impossible to do with a computer?

By now you should be comfortable programming in Fortran so I will leave it to you to write a program that can numerically integrate a function between known bounds using the mid-ordinate rule. The improvement in accuracy should be on the order of h, written $\mathcal{O}(h)$. In other words if the subinterval width h is halved, i.e. N is doubled, then the error in the numerical approximation of the integral is also halved.

5.1.2 The Trapezoidal Rule

One immediate, and intuitive improvement, we can make to the mid-ordinate rule is to make our subinterval strips approximate the function between the subinterval limits rather than just us the mid-point value. The easiest way to do this is to approximate the function as a straight line between the values for the function at the subinterval limits. In essence, we make the rectangle a trapezoid.

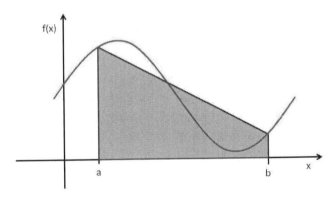

Figure 5.2: Illustration of the primitive trapezoidal rule

If we consider the entire interval as one strip, as illustrated in Figure 5.2, satisfy yourself that

$$\int_a^b f(x)dx \approx (b-a)\frac{f(a)+f(b)}{2}. \tag{5.3}$$

Equation (5.3) is referred to as the primitive integral. Subdividing this into N strips we obtain

$$\int_a^b f(x)dx \approx h\frac{f(a)+f(b)}{2}+h\sum_{n=1}^{N-1}f(x_n) \tag{5.4}$$

where $h = (b - a)/N$, and $x_n = a + nh$. Equation (5.4) is referred to as the composite (trapezoidal) integral.

The trapezoidal rule should have an error reduction that is proportional to $\mathcal{O}(h^2)$; halve h and you reduce the error by a factor of four. To confirm this write a Fortran program that performs the composite trapezium rule on a function you can integrate analytically.

5.1.3 Simpson's Rule

As a next step in improving the accuracy of our numerical integration scheme we might consider approximating the integrand as a piecewise quadratic. This is exactly what Simpson's Rule does and is illustrated in Figure 5.3. The derivation of Simpsons Rule involves taking a Taylor series expansion about the mid-point of the interval, and integrating that expansion. I will leave it as exercise to the reader to derive Simpsons rule in this manner. The formula for the primitive Simpsons Rule, i.e. the whole interval taken as one strip, is

$$\int_a^b f(x)dx \approx (b-a)\frac{f(a)+4f(c)+f(b)}{6} \tag{5.5}$$

where c is the mid-point of the interval. We need three function evaluations in this case because we are approximating the function with a quadratic that requires a minimum of three points. The composite Simpsons Rule has the form

$$\int_a^b f(x)dx \approx \frac{h}{3}[f(a)+f(b)+4\sum_{n=1}^{N/2} f(x_{2n-1})+2\sum_{n=1}^{(N/2)-1} f(x_{2n})] \quad (5.6)$$

where $h = (b-a)/N$ and $x_n = a + nh$. Again write a program that performs the composite Simpson's Rule for numerical integration and investigate how the error behaves in terms of the strip width. Note that you will need the total number of strips, N, to be an even number.

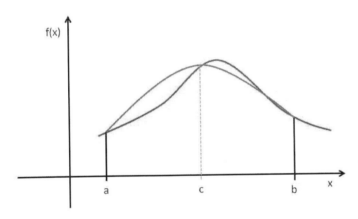

Figure 5.3: Illustration of Simpson's rule. Here the definite integral of the function (red curve) is approximated by area under the quadratic (blue curve)

5.2. Advanced Quadrature

5.2.1 Euler-McClaurin Integration

We could of course keep going with the approximations to the integrand function using higher ordered polynomials. Indeed using a cubic polynomial we are lead to Simpson's three-eighths rule and using a quartic polynomial yields Boole's rule but these soon become very cumbersome to derive and use. Is there a method to keep the relative simplicity of the trapezoid rule but with better accuracy? The answer is yes, I wouldn't have asked the question otherwise. The integration scheme to which I am referring is called the Euler-McClaurin scheme, given by the composite formula

$$\int_a^b f(x)dx = h\left(\frac{f(a)+f(b)}{2}\right) + h\sum_{n=1}^{N-1} f(x_n) + \frac{h^2}{12}[f'(a)-f'(b)] - \frac{h^4}{720}[f'''(a)-f'''(b)] + \cdots. \quad (5.7)$$

Equation (5.7) can be derived by again considering the Taylor series expansion of the function and its derivatives at the integration limits. See ref[DeVries] pages 153–155 for a neat explanation of the derivation. The first two terms in equation (5.7) are simply the trapezoid rule and we can consider the next terms as corrections to that numerical integration scheme. Note that should the first order derivatives at the integration limits be near identical, or vanishingly small, the trapezoid rule can give surprisingly accurate results. Can you think of a function whereby its derivatives will be *identical* for a given set of integration limits?

Although the Euler-McClaurin formula is far superior to any other numerical integration method we have discussed so far it suffers from the drawback that the integrand has to be easily differentiable. If not we would have to rely on numerical approximations of the derivatives at the integration limits. The accuracy of those derivative approximations should at least match the order of h to which they belong. This quickly becomes impractical.

5.2.2 Adaptive quadrature

In the previous discussions it is assumed that the strip width is uniform across the integration interval. To those experienced in numerical integration this is wasting considerable effort. Typically we want to determine the value of a numerical integration to some predetermined accuracy or tolerance. With our current numerical integration schemes we can only reduce the error by reducing the size of each strip. For smooth functions this is fine; the contribution of each strip to the total absolute error is roughly the same. However, what if the function is not smooth, or has portions that rapidly change with x, compared to other flat regions. For instance consider the Lorentzian line-shape function that describes the emission of light from the atoms of an excited gas cloud

$$I(\lambda) = \frac{I_0}{1 + 4(\lambda - \lambda_0)^2 / \Gamma^2} \tag{5.8}$$

where λ is the wavelength of light emitted, λ_0 is the resonant wavelength, I_0 is the peak intensity of emitted light at $\lambda = \lambda_0$, and Γ is a measure of the width of the curve; full width at half height.

This function is sketched in Figure 5.14; note that a small constant background intensity has been included. Let's say we are performing an experiment to determine how the width of the peak is affected by the pressure of the gas. Being good scientists we want to ensure that the total number of contained atoms remains constant at each measured pressure level, within some predetermined tolerance of say 0.1%. One way to do this would be to check the total emitted power of the gas at each pressure, i.e. the area under the curve in Figure 5.4. This means integrating the line-shape over some predetermined wavelength range. Let's assume we don't know how to integrate this type of function analytically[*] and so we have to do it numerically. The relative error in our numerical approximation should be at least equal to tolerance we want for our total emitted power measurements and ideally much less, let's say 0.001%. The

---

[*] It is actually a standard integral after a simple substitution

majority of the error in the approximation will be introduced by those strips representing the peak, and we require relatively narrow strips in this region order to keep the overall error below what we are willing to tolerate. Using a uniform distribution of strips we would be wasting effort in the relatively flat regions away from the peak; each strip would produce an insignificant portion to the overall error and we could afford to use wider strips in these regions.

A first approach to this problem might be to manually segment the integration interval into three subintervals; the two flat regions, and the peak. The strips for each subinterval could be chosen so that the absolute error of each was one third that required of the total. This approach is perfectly valid but hardly provides a general solution; where do you select the segmentations? What if there is still significant function variation within the selected subinterval? Would greater manual segmentation require more or less effort?

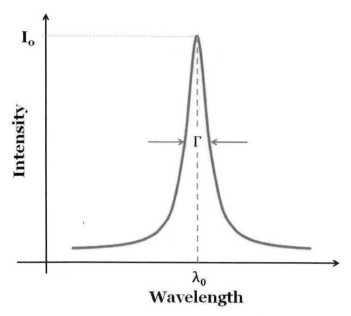

Figure 5.4: Sketch of the Lorentzian line shape.

To provide a more general solution let's consider the behaviour of the error of the trapezoid rule. We know that if you halve the strip width then you reduce the error by a factor of four, in other words the trapezoid rule has an error behaviour of $\mathcal{O}(h^2)$. If we denote the trapezoidal approximation using 2^m strips as T_m, then the trapezoidal approximation with twice the number of strips (half the strip width) is given by T_{m+1}. As T_{m+1} has half the strip width its error is reduced by a factor of four compared with T_m. With reference to the Euler-McClaurin equation (5.7) we can eliminate the leading error term in our approximation by performing the following calculation

$$T'_{m+1} = \frac{4T_{m+1} - T_m}{3} \tag{5.9}$$

where T'_{m+1} is the improved approximation for 2^{m+1} strips. If it helps you can think of equation (5.9) as a weighted average between the two approximations T_m and T_{m+1}. As we have eliminated the leading error term from equation (5.7) the error in our improved approximation is now $\mathcal{O}(h^4)$. It is worth noting here that Equation (5.9) is equivalent to Simpson's rule. We can now estimate the error in the T'_{m+1} by subtracting T_{m+1} giving

$$\varepsilon \approx T'_{m+1} - T_{m+1} = \frac{T_{m+1} - T_m}{3}. \tag{5.10}$$

Note that this is the *absolute* error **not** relative error. We can now check this estimated value against the "global" error we want to achieve in our approximation. If this condition is met we accept the integration, if not then we halve the total integration interval and perform the same process on the two halves. Note that the "global" error needs to be halved for these new subintervals so as to preserve the global error when they are summed for the entire integration. This procedure is repeated until the desired accuracy is reached upon which we accept the integration for that subinterval, add it to the total, and move on to the next subinterval.

Although not obvious at all this problem is best suited to a recursive function or subroutine whereby each successive call halves the subinterval and the "global" error that we check against. The recursion is terminated once the error estimate of equation (5.10) becomes sufficiently small. Getting your head around the logic of recursive formulas can be quite tricky and I find the best way to understand them is to visualise some simple output.

The program a*daptiveQuad.f90* performs this recursive action using the trapezoid rule with the improvement technique described as above. Figure 5.5 shows the result of applying this program to equation (5.8) with $\lambda_0 = 10$, $I_0 = 1$, $\Gamma = 1$, and with integration limits of $a = 6$ and $b = 14$. The global error was selected to be 0.01% of the integration. Here we can see that the adaptive quadrature has done its job; in the flat regions away from the peak the strips are wider, whereas the strips covering the peak are much narrower. Note that here the strips plotted belong to the T_0 integration in order to lend clarity to the figure; the strips used in the actual calculation of the integration are half the size of the ones shown. To visualise the data in such a manner the limits of the subintervals used are stored along with their function evaluations and plotted as impulse lines on the same figure as the function using 'gnu plot'.

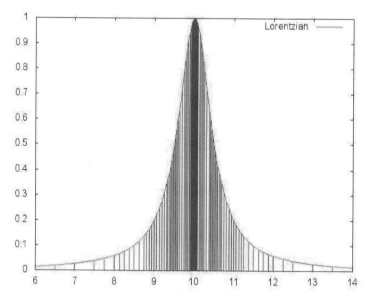

Figure 5.5: Adaptive numerical integration of the Lorentzian lineshape function

I alluded to the fact that the Lorentzian line-shape function can be reformed into a standard integral using a simple substitution. That substitution is

$$x = \frac{2(\lambda - \lambda_0)}{\Gamma} \quad (5.11)$$

which gives the integration of the (normalised) line-function the following form

$$\frac{1}{I_0}\int_a^b I(\lambda)d\lambda = \frac{\Gamma}{2}\int_c^d \frac{1}{1+x^2}dx \quad (5.12)$$

where c and d are the adjusted integral limits after the substitution of equation (5.11). Equation (5.12) is a standard integral that has the following exact solution

$$\int \frac{1}{1+x^2}dx = \arctan(x). \quad (5.13)$$

This analytical solution allows us to check the validity of our numerical solution and that it satisfies the requirement that the global error is at or around 0.01%.

5.2.3 Multidimensional integration

Multidimensional integrations pop up often in physics and generally require much more effort to solve than the one dimensional case. Take for instance a two dimensional integral of the form

$$I = \int_a^b \int_c^d f(x, y) dx dy \quad (5.14)$$

where $f(x, y)$ is a two dimensional function, and x and y have their usual Cartesian meaning. Here the integration limits a, b, c, and d specify a region in the x-y plane. With one-dimensional integration we are finding the *area* between the function curve and the x-axis bounded by the given limits. Similarly, two-dimensional integration

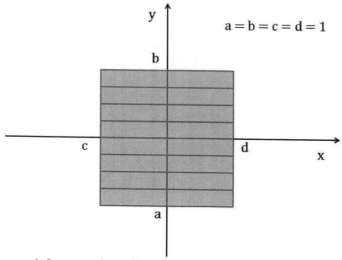

Figure 5.6: Square region split into strips running parallel to the x-axis. Here we do the x integration first.

finds the *volume* between the function *surface* and the x-y plane bounded by an area or region. In three dimensions the integration finds a four-dimensional space bounded by a three-dimensional surface[*]. This increase in dimensionality can continue *ad infinitum* (or *ad nauseam* depending on your philosophical bent) in a mathematical sense but typically stops at three when considering most physical phenomena.

The general strategy in numerically solving a two-dimensional integration is to split it into strips along one of the dimensions and treat each strip as a one dimensional integration along the other dimension. The total volume of the integral is found by adding together the contributions from each strip. Figure 5.6 illustrates this point. Here we have a region in the x-y plane that is bounded by the unit square, centred at the origin; the integration limits of equation (5.14) are constants. The region has been broken into strips running

---

[*] If you're having trouble visualising this – well done, you're normal.

parallel to the x direction. The function axis is pointing out from the plane of the page, and the function itself will form some surface either above, below, or cutting through the plane of the page (x-y).

Mathematically, we have split the two-dimensional integration into two, nested, one-dimensional integrals such that

$$I = \int_a^b F(y)dy \qquad (5.15)$$

where

$$F(y) = \int_c^d f(x,y)dx . \qquad (5.16)$$

Of course we can always reverse the order of the integration so that the y variable is integrated first. This would be equivalent to having strips running parallel to the y-direction of the square. When it comes to writing code to perform two-dimensional integration we can go two ways; (1) have a subroutine with nested loops or (2) have two separate subroutines to perform the integration in each dimension. The choice is yours, however I recommend that you make use of the code(s) already developed so that you are not wasting effort.

In many real physical systems the region may not be a square but some other more complicated shape, where the limits of the integral in one dimension are a function of the limits in the other. Take for instance the unit circle located at the origin. This has the equation

$$y = \sqrt{1-x^2} \qquad (5.17)$$

If we integrate over the upper right quadrant of the circle we can see that the limits in the x integration have to adjust depending on the y value for which we are calculating. However, like many problems in physics we can take advantage of the symmetry of the system and a change to polar coordinates yields constant integration limits, as illustrated in Figure 5.7. The strips become concentric rings centred on the origin. It is likely this change of coordinates will make the integrand function more complex but why should we be concerned? We're performing numerical integration in the first place because the problem was too difficult/impossible to solve analytically.

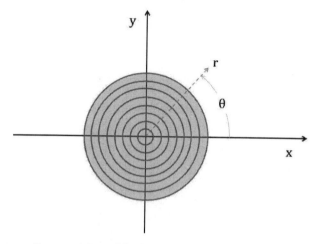

Figure 5.7: Segmentation of the (unit) circle region in polar coordinates

Exercises

1. Compare the effort required to find a numerical approximation of the integration

$$\int_0^1 f(x)dx = \int_0^1 x(1-x)dx$$

to an accuracy of 5 significant figures using the various methods we have developed in this chapter.

2. The time period of a pendulum in confined to a single plane without damping has the following formula

$$T = 4\sqrt{\frac{l}{2g}} \int_0^{\theta_0} \frac{d\theta}{\sqrt{\cos\theta - \cos\theta_0}},$$

where l is the length of the pendulum, g is gravitational acceleration, θ is the angle between the pendulum and the vertical, and θ_0 is the initial angle of release. Calculate this integral numerically for various initial angles of release and thus establish what is meant by the small angle approximation.

3. Write a (sub)program that performs integration over two dimensions using a method of your choice. The adventurous amongst you might like to try the adaptive quadrature in two dimensions.

4. What value can you approximate by numerically integrating over the unit quarter circle? Find this value to 8 significant figures.

5. Consider a unit square region, centred on the origin, containing a uniform distribution of charge, ρ. The electrostatic potential at a point (x_m, y_m) outside this region is found by integrating over the charged region such that

$$\phi(x_m, y_m) = \frac{\rho}{4\pi\varepsilon_0} \int_{-1}^{1}\int_{-1}^{1} \frac{dxdy}{\sqrt{(x-x_m)^2 + (y-y_m)^2}}.$$

By taking $\rho = 4\pi\varepsilon_0$ and numerically assessing the integral at different points outside the unit square attempt to plot contour lines of isometric potentials. Do you recover Coulomb's law at distances far from the charged region?

6. The charge distribution in the previous question does not have to be uniform across the region. Try our different functions to see how they affect the electrostatic potential surrounding the square region.

6. Ordinary Differential Equations

"Little by little, one travels far."

- J. R. R. Tolkien

Physics is mostly concerned with phenomena that are in flux, for instance, things that change either in time or space or both, and many of the laws of physics are most conveniently formulated in terms of differential equations; formulas that relate derivatives to functions. As an example consider Newton's second law of motion for a particle of mass, m, in one dimensional motion under a force field $F(x)$:

$$F(x) = m\frac{d^2x}{dt^2}. \tag{6.1}$$

This is a second order differential equation as we are taking the second derivative of the displacement, x, called the dependent variable, with respect to time, t, called the independent variable. The force field $F(x)$ can be referred to as the derivative function.

Finding the numerical solution of differential equations is one of the most common tasks in computational physics as many of these equations become analytically insoluble (or at least very difficult to solve) when you include realistic physical processes.

Before we dive into the various numerical solutions of differential equations I just want to ensure that we have an understanding of the different classifications of differential equations that exist. Otherwise, you'll read a sentence like, "the following differential equations is a partial, non-linear, nonhomogeneous equation of the third order", and be left scratching your had saying "What the fudge?".

6.1. Classification of Differential Equations

6.1.1. Types of Differential Equations

Differential equations can be categorised into two major groups, ordinary differential equations (ODE) and partial differential equations (PDE). The difference between the two is that ODEs only have one independent variable (they still can have any number of dependent variables) and PDEs can have any number of independent variables as well any number of dependent functions. Simple harmonic motion (SHM) in one dimension is an example of an ODE:

$$m\frac{d^2x}{dt^2} = -kx, \tag{6.2}$$

where m is the mass of the body in motion, k is the so-called spring constant, and x represent the displacement from some equilibrium position. Here time, t, is the only independent variable and x is the dependent variable that is a function of the independent variable, normally written as $x(t)$. Whereas, the wave equation, which consists of second order derivatives in both space and time, i.e. two independent variables, is an example of a PDE[*]:

$$\frac{\partial^2 u}{\partial t^2} = c^2 \frac{\partial^2 u}{\partial x^2}, \tag{6.3}$$

where c is the speed of the wave, and u represents some (scalar) property of the wave (e.g. displacement, pressure, electric field strength, and so on); u is the dependant function in this case, normally written as $u(x, t)$. Note the use of the partial derivative symbol, ∂, rather than the usual d.

We can further subdivide the groups into their order. The order refers to the highest derivative appearing in the equations. For

---

[*] PDEs also come in different flavours including parabolic, hyperbolic, and elliptic; the wave equation is an example of an hyperbolic PDE.

example, Equation (6.2) is of order 2, as is Equation (6.3). Order 2 ODEs (and PDEs) occur very often in physics. Note that we can separate a second order ODE into a pair of coupled, first order ODEs should we so wish but more on that later.

Next we can classify a differential equation as being linear or non-linear. In a linear ODE the dependent variable and its derivatives only appear to the first power and are not cross multiplied. Note that the independent variable can be to any power. For instance

$$f'' = -f + x^3, \qquad (6.4)$$

is linear while

$$f'' = -f^2 + x^3, \qquad (6.5)$$

and

$$f''f' = -f + x^3, \qquad (6.6)$$

are non-linear as they contain the terms f^2 and $f''f'$ respectively. Here I have used the notation that

$$f' = f'(x) \equiv \frac{df(x)}{dx}, \qquad (6.7)$$

i.e. a dashed derivative is one taken with respect to a spatial variable, in this case x. A dotted derivative is one taken with respect to a temporal variable, usually the time t, thus

$$\dot{f} = \dot{f}(t) \equiv \frac{df(t)}{dt}. \qquad (6.8)$$

This is a generally accepted notation convention within physics and mathematics, and other sciences also.

A further classification can be made to distinguish between homogeneous and nonhomogeneous differential equations*. A homogeneous equation contains terms that include either the dependant variable or its derivatives, but no other function of the independent variable. The differential equation of the simple harmonic oscillator, Equation (6.2), is an example of a homogeneous equation. Adding a time dependant driving force, $F(t)$, to this equation gives the nonhomogeneous equation

$$m\frac{d^2 x(t)}{dt^2} + kx(t) = F(t), \qquad (6.9)$$

as we now have a function of the independent variable, t, on the right hand side. The actual form of the driving force is unimportant to this discussion but will be particular to the system being described by the differential equation. Note that Equations (6.4) to (6.6) are all nonhomogeneous due to the addition of the independent variable term on the right hand side.

6.1.2. Types of Solution and Initial Conditions

Now that we have established the various classes of differential equations I'd like to include a short section about distinguishing their solutions. When solving differential equations we draw a difference between the general solution and a particular solution. The general solution refers to all the functions that fit the differential equation, whereas the particular solution is defined by some initial conditions or values. Take for instance Newton's law of cooling (or heating) that states that the rate of change of temperature of a body is in proportion to the temperature difference between it and that of the ambient. It can be written in the form

---

*As an aside try to find out why they are called nonhomogeneous rather than heterogeneous.

$$\dot{y} = -ky, \qquad (6.10)$$

where y is the temperature *difference* between the body and the ambient, and k is some constant of proportionality (related to the surface area of the body, the material the body is made from, and so on). The minus sign represents the physics that hot bodies cool and cold bodies warm. This has the general solution

$$y(t) = y_0 e^{-kt}, \qquad (6.11)$$

where y_0 is the initial temperature difference, i.e. the temperature difference at $t = 0$. Given an initial temperature difference we could then determine a particular solution for any value of k. Note that y_0 is called an integrating constant that is a mathematical concept related to some initial condition of the system, whereas k is a constant purely related to the physics of the system; the two should not be confused.

Equation (6.10) is an ODE of the first order and only has one integrating constant. As such we only needed to know one initial condition to determine a particular solution, namely the initial temperature difference. In the general case, an n ordered ODE will produce n integrating constants and we need as many initial conditions in order to find a particular solution. To convince yourself of this what initial conditions are required in order to determine a particular solution of the second order ODE describing SHM?

These are known as initial value problems. Other integration constants may include boundary conditions i.e. the condition or value of the dependant function at or beyond the boundaries or edges of your modelled system. For instance, the physical properties of the potential barriers in a quantum well provide the boundary conditions for solving Schrodinger's wave equation for an electron trapped in the well. Essentially the wave function, and it's derivatives, decay to zero as it penetrates deeper into the bounding potential barriers (for the case where the potential barriers are of finite height).

6.2. Solving 1st Order ODEs

6.2.1. The Simple Euler Method

Leonhard Euler was an eighteenth century, Swiss born mathematician, who we would describe today as a polymath. Euler worked in several areas including optics, astronomy, ship construction, and artillery but was most prolific in his work on mathematics. He contributed much to the fields of number theory, algebra, and calculus, and can be directly attributed to the modern standard usage of the symbols e, π, and i. Although I could on at length about his accomplishments our present interests are in his methods for numerically solving differential equations.

Consider again Equation (6.10) which is a linear, homogeneous ODE of the first order. We can write a first ordered ODE more generally as

$$\dot{y}(t) = f(t,y), \tag{6.12}$$

where f is some function of the independent variable, t, and the dependent variable, y; the form of f determines the classification of the differential equation.

We could attempted to solve Equation (6.12) by taking the Taylor series expansion of the dependent variable about some initial position, t_0, such that

$$y(t) = y(t_0) + (t-t_0)\dot{y}(t_0) + \frac{(t-t_0)^2}{2!}\ddot{y}(t_0) + \cdots. \tag{6.13}$$

Since we know the form of the first ordered derivative from Equation (6.12) we could calculate the higher ordered derivatives by using the partial differentiation of f. However, this soon becomes untenable for all but the most simple expressions for f, and in which cases Equation (6.12) can most likely be solved analytically. We can get rid of those troublesome higher order derivatives by truncating Equation (6.13) to the first two terms only, leaving

$$y(t) \approx y(t_0) + (t - t_0)\dot{y}(t_0). \tag{6.14}$$

Note that we could have arrived at Equation (6.14) by considering an approximation to the gradient of the dependent function at the initial position

$$\dot{y}(t_0) \approx \frac{y(t) - y(t_0)}{t - t_0}. \tag{6.15}$$

If we now say that $h = (t - t_0)$ where h is a small step, we may now conveniently write Equation (6.14) as an equality

$$y(t_0 + h) = y(t_0) + hf(t_0, y(t_0)) = y_0 + hf_0, \tag{6.16}$$

where we have substituted in the function, f, for the derivative, and used the notation that $y_0 \equiv y(t_0)$ and $f_0 \equiv f(t_0, y_0)$.

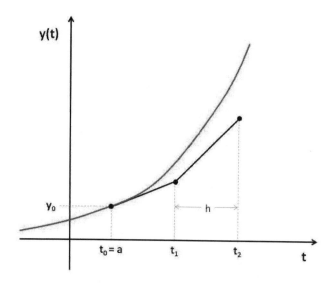

Figure 6.1: Sketch of the simple Euler method. We only know the first value of *y* exactly; the integrated values are approximations to *y*.

Equation (6.16) is the simple Euler method or Euler's forward approximation. Interpreting this method we can see that given a starting position, y_0, we can step to the next position using the derivative f_0 at the start of the step. We can generalise this to the nth step giving the recursion formula*

$$y_{n+1} = y_n + hf_n, \tag{6.17}$$

where we define, $t_n \equiv t_0 + nh$, $y_n \equiv y(t_n)$, and $f_n \equiv f(t_n, y_n)$, with $n = 1, 2, 3 \cdots$. Note that this stepping action can be referred to as integrating the solution; we are solving a differential equation and are therefore performing an integration. Figure 6.1 illustrates the simple Euler method in action.

Although we could step *ad infinitum* typically we wish to find a value for y at some predefined value for t, in other words, we have an interval $[a, b]$ over which we wish to step from a to b. The most straight forward way of doing this is to split the interval into N steps of equal size, h, such that $h = (b - a)/N$, and move the solution along one step at a time using Equation (6.17). We can check the accuracy of the method by repeating the integration using smaller and smaller step sizes and seeing if we converge on a solution. Though perhaps a more stringent test is that once we reached our desired value b we integrate *back* towards a and compare our integrated approximation to the initial value for y with which we started.

As we arrived at Equation (6.17) using a truncated Taylor series expansion we can determine the behaviour of the error in our approximation to the solution. For a *single step* of the simple Euler method we know that the approximation given by Equation (6.14) will have an upper bound on the "local" error of $\mathcal{O}(h^2)$ as we kept the first two terms of the Taylor series only. To get from our initial position a to our desired position b we have to perform N such steps. Thus the overall upper bound on the error at b will be given by $N \times \mathcal{O}(h^2)$. As N is inversely proportional to the step size h, we can then estimate the "global" error in the simple Euler method as $\mathcal{O}(h)$. In other words if you halve the step size (and thus take twice as many steps) you should halve the error in the approximation to the solution at the destination b.

---

* So called because you repeat the same calculations at each step to get to the next step.

The program *euler1.f90* performs the simple Euler method on the differential equation

$$y' = -xy, \qquad (6.18)$$

with the initial condition that $y_0 = 1$, and over the interval $x = [0,2]$. The analytical solution to Equation (6.18) with the given initial condition is

$$y = e^{-0.5x^2}. \qquad (6.19)$$

In its current state the program performs the integration initially with $N = 5$, printing to file ('euler1a_data_<N>.txt') the values of x, the approximation y, and the analytical solution at each step. The program then reports to screen the step size h that was used and the global error achieved by the approximation before increasing N by 5 so that $N = 10$, and repeating the process. This continues up to a maximum of $N = 20$. Compiling and running this program you should observe the $\mathcal{O}(h)$ error behaviour we predicted.

6.2.2. Modified and Improved Euler Methods

Although the simple Euler method provides an instructive means of introducing the topic of numerically solving differential equations it should not be used in any serious attempt to find a solution due to its lack of accuracy. The major issue with the simple Euler method is that it assumes the derivative at the start of a step remains constant over that step, see Figure 6.1. This asymmetrical treatment of the step is bound to lead to large inaccuracies of the approximated solution. It would be better if we could use some sort of averaged value to estimate the derivative across the whole step.

The modified Euler method approximates the solution by using the derivative at the mid-point of the step to advance the integration. Obviously, we don't know the value of the derivative at the mid-point but we can approximate it using the simple Euler method with half the step size such that

$$t_{mid} = t_{n+\frac{1}{2}} = t_n + \frac{h}{2} \tag{6.20}$$

and

$$y_{mid} = y(t_{mid}) = y_{n+\frac{1}{2}} = y_n + \frac{h}{2} f_n, \tag{6.21}$$

where n is our previous step for which we have values. We can now use the value for y_{mid} to estimate the derivative at the mid-point of the step and thus advance the solution across the whole step as follows

$$y_{n+1} = y_n + hf(t_{mid}, y_{mid}). \tag{6.22}$$

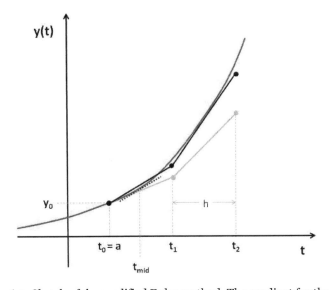

Figure 6.2: Sketch of the modified Euler method. The gradient for the entire step is estimated from the derivative at the mid-point. The simple Euler method is shown in grey for comparison

Equation (6.22) is the modified Euler method and is illustrated in Figure 6.2. From a cursory look at the figure you can see that the modified Euler method appears to do a much better job at approximating the solution than the simple Euler method*.

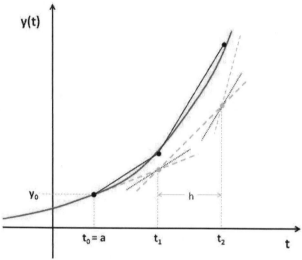

Figure 6.3: Sketch of the improved Euler method. The simple Euler method is used to estimate the derivative at the end of the step, which combined with the derivative at the start of the step gives a mean for the entire step.

Another way of obtaining an average value that best approximates the derivative across the step is to take a mean of the derivative at the start of the step with the derivative at the end of the step. Again we use the simple Euler method but this time to estimate the value of the derivative at the *end* of the step. Using this estimate with the derivative at the start of the step, which we have previously computed, we can take the mean of these two values to advance the integration one whole step. Mathematically we write

$$y_{n+1} = y_n + \frac{h}{2}[f_n + f(t_n + h, y_n + hf_n)]. \tag{6.23}$$

Equation (6.23) is the improved Euler method and is illustrated in Figure 6.3. Again a cursory study of the figure suggests that the

---

* Note that the function sketched is somewhat arbitrary – you should however be able to show that he modified Euler method is better by writing a program.

improved Euler method is better than the simple Euler method. But which is better between the improved Euler method or the modified Euler method?*

Use the simple Euler method program provided to help you write code for the modified, and improved Euler methods; Equations (6.22) and (6.23) respectively. Using your programs can you determine how the error behaves with step size for these two methods?

It was the German mathematician Karl Heun who first developed the modified Euler and improved Euler methods, which in part helped develop the more accurate Runge-Kutta methods that I will discuss next.

6.2.3. The Runge-Kutta method

Carl Runge and the polish born Martin Kutta were both German mathematicians and physicist who lived and worked around the latter part of the nineteenth century and into the first half of the twentieth century. In 1901, they co-developed the Runge-Kutta method(s), used to solve ordinary differential equations numerically.

The Runge-Kutta methods are characterised by expressing the numerical approximation in terms of the derivative function evaluated at intermediary points between step values. Euler's methods can actually be classed as low ordered general Runge-Kutta methods; the Euler method being a one step Runge-Kutta method, and the modified, and improved methods are both two-step Runge-Kutta methods. The popularity of the Runge-Kutta methods in numerically solving ODEs is due in part to their (relative) ease of implementation within computer programs, and the accuracy they achieve. Of the most popular devised is the 4th ordered Runge-Kutta (RK4), or simply *the* Runge-Kutta method, defined as

---

* There's only one way to find out ... fight!! ...I mean write a program. Also it's not the only way to find out....still its mildly humorous, if you like Harry Hill.

$$y_{n+1} = y_n + \frac{h}{6}(k_0 + 2k_1 + 2k_2 + k_3), \qquad (6.24)$$

where

$$k_0 = f(t_n, y_n),$$

$$k_1 = f(t_n + \frac{h}{2}, y_n + \frac{h}{2}k_0),$$

$$k_2 = f(t_n + \frac{h}{2}, y_n + \frac{h}{2}k_1)$$

$$k_3 = f(t_n + \frac{h}{2}, y_n + hk_2). \qquad (6.25)$$

The derivation of Equations (6.24) and (6.25) is a little tricky but involves considering a general form for Euler's methods such that

$$y_{n+1} = y_n + h[\alpha f_n + \beta f(t_n + \gamma h, y_n + \delta h f_n)], \qquad (6.26)$$

and choosing the coefficients $(\alpha, \beta, \gamma, \delta)$ such that they agree with the Taylor series expansion of the term involving β, up to h^4. Indeed, the modified and improved Euler methods can be found in this way by matching terms up to h^2. Feel free to have a go at deriving these equations yourselves using this method but be warned that taking a Taylor series expansion of a function of two variables gets somewhat complex for anything more than the degree one terms.

A slightly easier but less general way of deriving Equations (6.24) and (6.25) is to consider the direct integration of Equation (6.12) for the special case that derivative function is a function of the independent variable alone, i.e. $f(t)$. We can then write for a single step

$$y(t_n + h) = y(t_n) + \int_{t_n}^{t_n + h} f(t) dt, \qquad (6.27)$$

and by solving the integration term by Simpson' rule we obtain the results of Equations (6.24) and (6.25).

Interpreting the RK4 approximation, Equation (6.24), we see that the next value in the integration is calculated as the present value plus the weighted average of four increments. The increments are determined from estimates of the slope at intermediary points on the step specified by the derivative function f, multiplied by the step size h. These increments can be described as follows:

- k_0 is the estimate of the slope at the beginning of the step;
- k_1 is the estimate of the slope at the midpoint, using k_0;
- k_2 is the estimate of the slope at the midpoint, but now using k_1;
- k_3 is the estimate of the slope at the end of the interval, using k_2.

In averaging the four increments, greater weight is given to the increments at the midpoint reflecting the fact that the function's slope is better approximated by the tangent to the curve at the midpoint of the interval rather than its bounds.

The RK4 method is a fourth-order method, meaning that the local error behaves as $O(h^5)$, while the global error behaves as $O(h^4)$. This means that halving the step size will reduce the overall error by a factor of sixteen, hence why the method is so popular.

The module code *rk4_mod.f90* contains the subroutine 'RK4' which performs the fourth ordered Runge-Kutta method. I have left it as a task for you to write a program that calls this subroutine to solve the differential equation that is also defined in the module. This is the same as Equation (6.18) found earlier in this chapter. The variables used in the subroutine should be self explanatory. To compile a program that uses the module you write at the command line something like

gfc rk4_mod.f90 rk4_prog.f90 −o rk4

where in this case 'rk4_prog.f90' is the file containing your program code and the output is written to the executable 'rk4'. Confirm that the Runge-Kutta method provides an accuracy of $O(h^4)$.

6.2.4. Adaptive Runge-Kutta

Generally speaking we wish to find the numerical solution to an ODE to some predefined (global) error or tolerance. Using a fixed step size we are somewhat constrained to use one sufficiently small across the entire interval to provide the required local accuracy at each step of the integration. If the nature of the solution changes across that interval, i.e. becomes increasingly rapid in its variation with the independent variable, then we would waste considerable effort over the 'flat' regions of the solution. Therefore, we would like to be able to change the size of the steps taken in the numerical approximation in accordance with the local nature of the solution, i.e. allow them to adapt.

By far the most straight forward way to do this is to perform a single step of the integration with step sizes h and $h/2$ and compare the result immediately. More precisely, we perform a single step with step size h, then halve its size and perform two steps with the new step size in order to reach the same point in the solution. We can estimate the error in the numerical approximation by computing the difference between our two solutions. By comparing this difference to our predefined tolerance we can either accept the step if the difference is smaller, otherwise we use the halved step size to repeat the process. However, this is *not* the whole picture. Here we have *only* taken account of the solution starting in a flat region and advancing into a rapidly changing one, but what about the other way round?

If we start in a rapidly changing region then we merrily halve our step size until it produces a solution that is within the prescribed accuracy tolerance and we advance with that small step. If the solution now flattens then we simply maintain that small step as it will definitely produce a solution that is (very much) within the accuracy tolerance. This is *not* what we were after; we want a step size that adapts to the local nature of the solution, i.e. can increase as well as decrease.

The answer to this issue is to assume that when we accept a step, i.e. we are within the accuracy tolerance, the step size is too small and should be increased for the next step; we should double the step size to $2h$.

In the module code *rk4_mod.f90* you should have found the subroutine 'RK4A' at the end of the file. This performs the adaptive step process that I have outlined above. At the point where we accept the step I have made use of the fact that we have two approximations to the solution made with two different step sizes, namely h and $h/2$. We know how the error behaves in the *fourth ordered* Runge-Kutta method for decreasing step size so can use this information to eliminate the leading error term. This is the same method used in Chapter 5 to develop an adaptive quadrature by using the knowledge of how the error in the trapezoidal rule behaves with strip width. Generally, this method of manipulating the solution based on error behaviour is referred to as Richardson's extrapolation that I explore further in the advanced section of this book.

6.3. Solving 2nd Ordered ODEs

6.3.1. Coupled 1st Order ODEs

It has been noted before that second order ODEs occur most frequently in physics as they model many real physical systems. In general we write a second order ODE as

$$\ddot{y} = f(t, y, \dot{y}). \tag{6.27}$$

Note that the function f has three variables namely the independent variable, the dependent variable, and the first derivative of the dependent variable.

Although methods exist to solve higher ordered differential equations, e.g. finite difference method, it is far simpler to reduce the equation into a set of coupled first order differential equations; the term coupled will become apparent shortly. We can do this by introducing secondary dependent functions such that $y_1 = y$ and $y_2 = \dot{y}$ and thus we can rewrite Equation (6.27) as a pair of coupled, first order ODEs:

$$\dot{y}_1 = y_2;$$
$$\dot{y}_2 = f(t, y_1, y_2). \tag{6.28}$$

They are coupled because the rate of change of variable y_1 is dependent on the variable y_2, and the rate of change of variable y_2 is dependent on the variable y_1 contained in the function f. At first glance you may think so what? It still looks like a nightmare to solve. However, if we define $f_1 \equiv y_2$ and $f_2 \equiv f(t, y_1, y_2)$ then Equations (6.28) can be rewritten in vector form thus

$$\begin{bmatrix} \dot{y}_1 \\ \dot{y}_2 \end{bmatrix} = \begin{bmatrix} y_2 \\ f(t, y_1, y_2) \end{bmatrix} = \begin{bmatrix} f_1 \\ f_2 \end{bmatrix}, \tag{6.29}$$

or

$$\underline{\dot{y}} = \underline{f}, \tag{6.30}$$

where $\underline{\dot{y}}$ and $\underline{f}$ represent two component vectors. Comparison of Equation (6.30) with Equation (6.18) shows that the problem of solving second ordered ODEs is not primarily different from the first ordered ODEs for which we have been developing solutions, only that now we have extra components.

To illustrate this point we can write Equation (6.1), which describes Newton's second law of motion, as a pair of coupled first order differential equation by introducing momentum as a secondary dependent variable. The momentum of a body of mass m in one dimension is defined as

$$p(t) \equiv mv(t) = m\dot{x}, \qquad (6.31)$$

where $v(t)$ is the velocity of the body at time t, and x represents the (one dimensional) displacement of the body in some coordinate system. Thus Equation (6.1) can be rewritten in terms of the momentum as follows

$$\dot{x} = \frac{p}{m}, \qquad (6.32)$$

$$\dot{p} = F(t, x, p/m), \qquad (6.33)$$

where time, position (displacement), and velocity have been included in the force term for completeness*. We can check these equations make sense by assessing the situation when no net force acts on the body, i.e. $F = 0$. This implies a constant momentum that in turn implies an unchanging velocity, and thus we recover Newton's first law of motion. Just to restate and reinforce our nomenclature, here we call time t the independent variable, and the position x and the velocity p/m are the (coupled) dependent variables.

---

* Note that here we assume that mass is a constant of motion

6.3.2. Oscillatory Motion

At the beginning of this Chapter we briefly discussed the second order differential equation describing simple harmonic motion (SHM). Using Equations (6.32) and (6.33) we can rewrite this as a pair of coupled first order ODEs

$$\dot{x} = \frac{p}{m} \qquad (6.34)$$

and

$$\dot{p} = -kx. \qquad (6.35)$$

We can make life easier for ourselves by rewriting these equations in terms of velocity, v, instead of momentum, p, such that

$$\dot{x} = v \qquad (6.36)$$

and

$$\dot{v} = -\frac{k}{m}x. \qquad (6.37)$$

We can make this change as, in this case, the mass is assumed to be a constant of motion and so we are not changing the physics of the system only our notation. Note that in this form Equation (6.37) has no multiplicative constant in front of the derivative. This is the general strategy you should employ when solving differential equations (numerically or analytically), ensuring all physical constants appear with the derivative function where possible.

The module code *euler2_mod.f90* contains a subroutine that implements the simple Euler method to numerically solve a general second ordered ODE that has been modified into a pair of coupled, first order ODEs. The subroutine contains two array variables of dimension 2, namely 'Y' and 'K0', that represent the coupled dependent variables, and the estimate of their derivatives at the start of the step, respectively. Here the variable 'K0' has been used to reinforce the idea that the simple Euler method is a one step Runge-Kutta method.

To illustrate their use when applied specifically to SHM: '$Y(1)$' represents the position variable x; '$Y(2)$' represents the velocity variable v; '$K0(1)$' is also the velocity variable v but representing the right-hand-side of Equation (6.36); and '$K0(2)$' is the derivative function ('DF') or the right-hand-side of Equation (6.37). Note that, as I have kept the code general, the argument 'X' in the subroutine represents the independent variable, <u>not</u> the displacement x in the discussion above.

After writing a program to call this subroutine, where we also define the derivative function 'DF' (Equation (6.37)), and using the initial conditions $x(0) = 1$, $v(0) = 0$, with $k/m = 1$, and 100 steps, we obtain the output plotted in Figure 6.4 up to a time of $t = 15$ seconds[*].

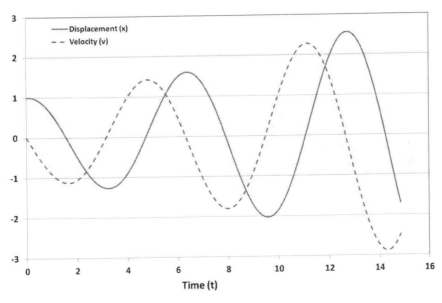

Figure 6.4: Numerical solution of SHM using the simple Euler method.

As the world's energy demand has yet to be satisfied by a mass-on-a-spring system what has gone wrong? Have we made a mistake with the physics or the implementation of the simple Euler method? To

---

[*] As we have kept everything else unit-less this could equally be any other measurement of time.

answer these questions let's examine the analytical solution of the ODE describing SHM. From your A-level or equivalent physics course you will know that the solution of the spring equation for displacement is a sinusoidal function in time. The phase of that solution, i.e. whether it is a sine function, cosine function, or somewhere inbetween, is dependent on the initial conditions. In the case above we obtain the particular solution

$$x = \cos(\omega t) \qquad (6.38)$$

where ω is the angular frequency of the oscillations, and is given by

$$\omega = \sqrt{\frac{k}{m}}. \qquad (6.39)$$

Thus we know that the solution is a cosine function with a (time) period of 2π. This is encouraging as our numerical solution, despite increasing in amplitude, has these properties, so it's a safe bet that the physics and the implementation are sound.

From our previous discussions of the Euler method it is fairly obvious that the instability of the numerical solution is down to the truncation error of the Taylor series. We could of course reduce the step size to help with the stability of the numerical solution but that would be wasting computational effort; we have already developed more accurate numerical solvers. Using the code 'euler2_mod.f90' as a template, add subroutines to perform the modified, and improved Euler methods, and see if they do any better solving the ODE describing SHM. As an aside, the results plotted in Figure 6.4 should make a strong case as to why the simple Euler method should not be used as serious attempt to solve ODEs describing real, physical systems. Indeed for our next discussion we require something with a bit more accuracy.

Returning now to our fixed step Runge-Kutta subroutine found in the module code *rk4_mod.f90*, I would like you to modify the code so that it can deal with a pair of coupled, first order ODEs. If you run into problems remember to use the code *euler2_mod.f90* as a guide; treat the Ks and the dependent variable Y as two component arrays, and include an auxiliary subroutine to deal with the derivatives.

Once you have a program up and running you should be able to reproduce the results plotted in Figure 6.5. Here we start with the same initial conditions and parameters as before but have allowed the integration to run up to time $t = 60$ seconds, and have changed the number of steps N to 600, i.e. a step length of 0.1 seconds.

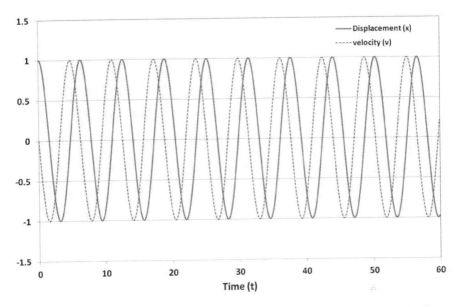

Figure 6.5: Fixed step Runge-Kutta solution of SHM, encompassing nine periods of oscillation.

Notice that the solution is stable for (at least) the first nine periods of oscillation. To stringently test the stability of the Runge-Kutta solution we should allow the integration to run over several thousand periods of oscillation, maintaining the same step length, and monitor the amplitude of the oscillations produced. However, the range of the stability shown in Figure 6.5 will be sufficient for the following discussion.

In real physical systems oscillatory motion is usually damped. We know this because after we put say, a mass on a spring in motion it will lose the initial amplitude it was given and eventually come to rest. This due mainly to resistive losses with the air it moves through. We can modelling this damping effect by assuming the drag force acting on the oscillating mass is proportional and opposite to the velocity of the mass. The second order ODE describing damped oscillatory motion now becomes

$$m\ddot{x} = -kx - D\dot{x}, \qquad (6.40)$$

where D is the constant of proportionality for the drag force, and I have taken the mass term to the left-hand-side for clarity. Note that Equation (6.49) remains a linear, homogeneous ODE of the second order. Separating Equation (6.40) into a pair of coupled first order ODEs is straight forward. In fact all we have to do is modify Equation (6.37) to include the additional, drag force term such that

$$m\dot{v} = -kx - Dv, \qquad (6.41)$$

and Equation (6.36) remains unchanged. This appears deceptively simple, and studying physics you will probably be developing a healthy mistrust of anything that appears simple, but in this case it is actually that straight forward.

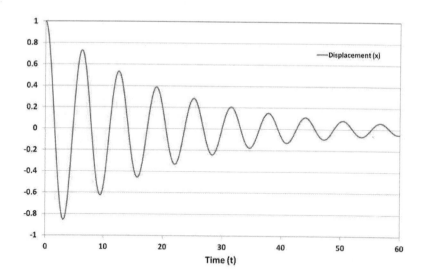

Figure 6.6: Damped oscillatory motion integrated using a fixed step Runge-Kutta method

After making the appropriate modification to the derivative function 'DF' in your code you should be able to reproduce the results plotted in Figure 6.6. Here I have taken the parameters to be $k/m = 1$ and $D/m = 0.1$. How would you deal with the drag force being proportional to the velocity squared instead?

As a last discussion to this section let's consider driven oscillatory motion. At the beginning of this Chapter I identified a nonhomogeneous ODE as one having a function of the independent variable extra to the dependent variable and its derivatives. I used driven oscillatory motion as an example in Equation (6.9) that I will repeat here with an addition of the drag term

$$m\dot{v} = -kx - Dv + F(t), \tag{6.42}$$

where $F(t)$ is the driving force. The form of the driving force will be dependent on the physics of the system Equation (6.42) describes. For instance, a child being pushed on swing[*] will have a driving force that would be well suited to being modelled an impulse acting at a particular point in the oscillation. Whereas a the driving force describing a car's suspension system as it travels over a cobbled street, say, could be modelled by some sort of sinusoidal function.

For the sake of this discussion and simplicity, let's assume the driving force is a straight forward sine function, thus

$$F(t) = A\sin(\omega_0 t + \phi), \tag{6.43}$$

Where A is the amplitude or maximum force supplied by F, ω_0 is the angular frequency of the driving force (the subscript identifies it from the angular frequency of the solution), and ϕ is a phase shift added for generality.

Make further modifications to your derivative function to include the driving force term as described in Equation (6.43). For now assume that the phase shift is zero. Keeping everything else the same and using the parameters $A = 0.2$ and $\omega_0 = 1$ you should be able to reproduce the results plotted in Figure 6.7.

---

[*] In this case we'd have to consider a pendulum system rather than a mass on a spring system.

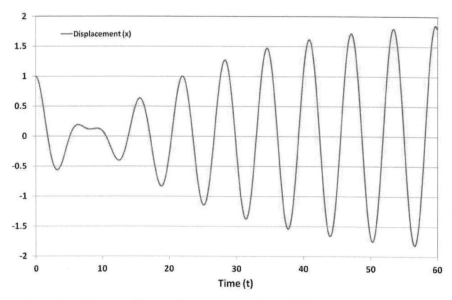

Figure 6.7: Driven oscillator. There are two regions of the solution: the transient region, and the steady-state region.

Notice that in the solution of the driven oscillator there are two regions. First is the transient region where the nature of the solution changes with the independent variable. Then the steady-state region where the oscillations follow the form of the driving force. Figure 6.7 shows the case where we have a situation close to resonance; though not actually at resonance - even though the driving frequency equals the natural frequency, the damping term effects the frequency at which the oscillator shows maximum response to the driving force (see Exercise 6).

6.3.3. More Than One Dimension

The code that we have developed to solve a second ordered ODE by transforming them into a pair of coupled first ordered ODEs have currently only considered motion in one-dimensional space. To solve for the motion we required two dependent variables, namely the displacement and the velocity (or momentum). These methods can be extended to cover problems involving several *dependent* variables that may describe motion in three-dimensional space.

For example, if we were to describe the motion of the Earth in orbit about the Sun, which we know is planar, we would need the Earth's x and y coordinates as well as it's velocities v_x and v_y. That is we require four variables to fully describe the Earth's orbit. For motion that isn't planar we would require six dependent variables to fully describe a bodies motion in four-dimensional space-time.

To modify the code to deal with these extensions is a relatively straight forward task. We simply increase the dimension of the dependent variable array 'Y' and the intermediary function evaluation arrays 'K' to the number we require, and increase the maximum loop index to the same number. Conventionally we put the dependent variables in first followed by their derivatives. For example, for the Earth's orbit about the Sun the dependent variables array will have following components:

$Y(1) = x$

$Y(2) = y$

$Y(3) = v_x$

$Y(4) = v_y$

The corresponding function evaluation arrays (found in the subroutine 'DERIV') will then be defined as

$K(1) = Y(3)$

$K(2) = Y(4)$

$K(3) = DF(X,Y,1)$

$K(4) = DF(X,Y,2)$

where we include an integer flag in the derivative function call to distinguish between the different dependent variable derivatives. Exercise 7 will require you to make these modifications.

Exercises

1. One stringent accuracy test of a numerical integration scheme is to have it step backwards from the value of the final step to the starting position and see how close we come to the initial value we supplied. Apply this test to the methods we have developed in this chapter and comment on the outcome for different step sizes.

2. Using Equation (6.18) and its solution (6.19) plot, on an appropriate graph, the error produced by the Euler methods and fixed step Runge-Kutta method for step sizes in the range $h = 0.1$ down to $h = 10^{-15}$. Comment on what you find.

3. Consider one-dimensional projectile motion with air resistance we can write
$$m\dot{v} = mg - Dv^2,$$
where m is the mass of the projectile, v is its velocity, g is the acceleration due to gravity, and D is the drag coefficient. For a sphere of mass $m = 10^{-1}$kg the drag coefficient was found to be $D = 10^{-3}$kg/m. Using one of the numerical solvers we have developed find the terminal velocity of the sphere dropped close to the surface of the Earth. Does this agree with theory?

4. Modify your Runge-Kutta program for SHM without any damping or driving force terms to check the stability of the method over a large number of periods (tens of thousands). How might you monitor the accuracy of the numerical solution?

5. Using your Runge-Kutta program for SHM with damping only, check for critical damping and assess when it occurs in terms of the relative values of k, m, and D. Does this agree with the theory of critical damping?

6. Using the numerical solvers we've developed, how does damping affect the resonance phenomena (resonant frequency and maximum response of the oscillator) of driven oscillations? Does this agree with real observations?

7. Newton's gravitational force of attraction between two objects is given by
$$\underline{F} = -\frac{GMm}{r^3}\underline{r},$$
where G is the universal gravitational constant, M and m are the masses of the two bodies, and r is their separation distance. Using either the fixed step or adaptive step Runge-Kutta method we've developed attempt to reverse engineer the mass of the Sun knowing that Earth requires one year to make the orbit. The distance between the Earth and the Sun is one astronomical unit (1 AU). Assume Earth's orbit is circular, that there is no influence from any other galactic body, and that the coordinates of the Sun are fixed at the origin. Try to get your answer to within 4 significant figures of accuracy (either check online or perform the calculation using the equations for circular motion).

7 Fourier Analysis

"The greatest challenge to any thinker is stating the problem in a way that will allow a solution."

- Bertrand Russell

Fourier analysis also known as spectral analysis is a powerful tool to the experimental scientist. It can help to establish a clear physical picture of an experimental system than just from the raw data on its own. Fourier Analysis can also be used to help extract significant information from particularly noisy or complicated signal or waveform that may have otherwise been missed or lost. For instance Fourier analysis can be used to: reconstruct a crystal structure from its X-ray diffraction pattern; determine the mass of ions exhibiting cyclotron motion in a magnetic field; reconstruct the 3-D image from a series of X-ray images in a Computerised Tomography (CT) scan; produce bandpass filters in electronic circuits; improve digital radio reception; clean up noisy digital images; and the list goes on. In general all these techniques rely on finding the Fourier transform of the measured, raw data. To do this we must first talk about how to represent or approximate a function using a Fourier series. As a starting point lets return to the Taylor series expansion of a function and discuss its limitations.

We saw from Chapter 2 that the Taylor series expansion is a powerful tool when it comes to approximating functions. However, as we saw with the truncated Taylor series expansion of the sine function it can only approximate a function reasonably accurately about the (unique) point it was taken. While this may not cause a significant limitation to most continuous functions, periodic functions are not well suited to Taylor series expansions; the period is simply not taken into account. In addition to this limitation with periodic functions the Taylor series expansion requires that a function and all its derivatives exist everywhere. In other words, the Taylor series expansion cannot be used to approximate functions with discontinuities (jumps) either in the function, or in the derivatives of the function. This is where Fourier steps in.

7.1 The Fourier Series

Jean Baptiste Joseph Fourier was a French mathematician and physicist born in Auxerre in 1768. He is best known for starting the investigation of the now eponymous Fourier series, that he applied to the then unsolved (general) problems of the propagation of heat and vibrations. It was Fourier who first pointed out that an arbitrary periodic function $f(t)$, with a period T, can be separated into a summation of simple trigonometric terms such that

$$f(t) = \frac{a_0}{2} + \sum_{n=1}^{\infty} \left(a_n \cos(n\omega_0 t) + b_n \sin(n\omega_0 t)\right), \tag{7.1}$$

where the a_n and b_n are the so called Fourier coefficients, and $\omega_0 = 2\pi/T$ is the natural frequency of the function. Note that *every* periodic function has a natural frequency *by definition* but only harmonic oscillators behave as pure sinusoidal waves. Interpreting the Fourier series we see that the function, which may represent some audio signal or EM radiation or whatever, is composed of the superposition of many harmonic tones of the natural frequency. A harmonic tone is a sinusoidal function with a period equal to an integer multiple of the natural frequency*. The coefficients a_n and b_n thus provide a measure of the contribution to the signal from the cosine and sine harmonics, respectively. More precisely, the intensity or power at each harmonic frequency is proportional to $a_n^2 + b_n^2$; this is referred to as the Fourier (power) spectrum.

The coefficients of the Fourier series are given by

$$a_n = \frac{2}{T} \int_0^T f(t) \cos(n\omega_0 t) dt, \tag{7.2}$$

and

$$b_n = \frac{2}{T} \int_0^T f(t) \sin(n\omega_0 t) dt. \tag{7.3}$$

---

* As a comparison with standing waves, you may think of the natural frequency as the fundamental frequency, with the integer multiples providing the overtones or harmonics.

Equations (7.2) and (7.3) can be derived directly from the Fourier series which I will leave as an exercise for the reader to perform (or research whichever is easiest). Note that we integrate over one period.

Equation (7.1) need not be restricted to periodic functions as any general function may be described by an infinite sum of its Fourier components (this is actually its Fourier transform which we discuss in the next section). Moreover, as this series does not require the derivatives of the function to exist it can be used to describe functions that are discontinuous, or contain discontinuous derivatives. We can think of the Fourier series as providing a "best-fit" to the function (or signal) in the least squares sense and it generally converges to the average behaviour of the function. At discontinuities it converges to the mean value of the function just either side of the jump, and at sharp corners, i.e. where there are discontinuities in the function's derivative(s), it overshoots the function.

So far this discussion has been somewhat abstract so let's go through an illustrative example. A square wave can be thought of as (periodic) repetition of a step function. A step function over a period T is given by

$$f(t) = \begin{cases} -A, & -\frac{T}{2} < t < 0 \\ A, & 0 < t < \frac{T}{2} \end{cases}, \qquad (7.4)$$

where A is the amplitude of the square wave. Given this definition the square wave is an odd function, i.e. $f(-t) = -f(t)$, and all the a_n must be zero; remember that the cosine is an even function i.e. $f(-t) = f(t)$, thus the integration of Equation (7.2) goes to zero over one period. Our job then is to find the b_n as follows

$$b_n = \frac{2}{T}\left(\int_{-T/2}^{0}(-A)\sin(n\omega_0 t)dt + \int_{0}^{T/2}(A)\sin(n\omega_0 t)dt\right)$$

$$= \frac{4A}{T}\int_{0}^{T/2}\sin(n\omega_0 t)dt$$

$$= \frac{4A}{n\omega_0 T}[-\cos(n\omega_0 t)]_{0}^{T/2}$$

$$= \frac{2A}{n\pi}[1-\cos(n\pi)]$$

$$= \begin{cases} 0, & n = 2,4,6... \\ \dfrac{4A}{n\pi}, & n = 1,3,5... \end{cases}$$

where we have use the fact that the function is odd and the natural frequency is defined as $\omega_0 \equiv 2\pi/T$. Substituting these values into Equation (7.1) and summing to infinity we would end up with the square wave. In practice, we typically only retain the first few significant terms from the Fourier series.

Figure 7.1 plots the result of performing the Fourier series for a square wave with an amplitude $A = 1$ and a period $T = 4$ (seconds). The plot shows the first three non-zero Fourier terms of the series namely $n = 1$, 3 and 5. As more terms are added we can see that the series does an increasingly better job of approximating the function. Where the square wave is a constant the Fourier series oscillates around the function value with decreasing amplitude as we increase the number of terms in the series. As the Fourier series passes through the discontinuity in the function it converges on the average value of the function limits either side of the jump (in this case zero) and misses the function entirely just passed the jump. This is the *overshoot* that I mentioned previously. Unlike the oscillations about the constant function value (more generally the continuous parts of the function) the overshoot does not improve as we increase the number of terms in the series.

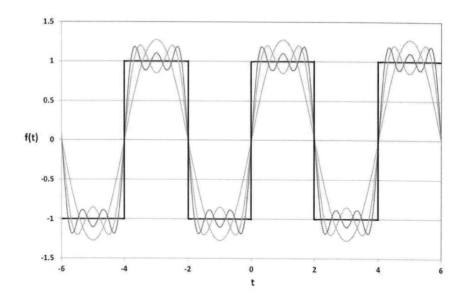

Figure 7.1: Approximation of the square wave using a Fourier series, keeping the first, second, and third non-zero terms in the series.

The majority of the problems and functions that we'll deal with will involve real numbers, that is to say, not complex numbers. However, it is sometimes convenient to express the Fourier series in terms of complex numbers. Returning briefly to Euler who derived the following complex identities

$$e^{i\theta} = \cos(\theta) + i\sin(\theta) \qquad (7.5)$$

and

$$e^{-i\theta} = \cos(\theta) - i\sin(\theta) \qquad (7.6)$$

where $i = \sqrt{-1}$ is the so-called imaginary number we can rewrite the Fourier series as

$$f(t) = \sum_{n=-\infty}^{\infty} c_n e^{in\omega_0 t}. \qquad (7.7)$$

The coefficients are now represented by the c_n which can be calculated using

$$c_n = \frac{1}{T}\int_0^T f(t)e^{-in\omega_0 t}\,dt. \qquad (7.8)$$

Before moving on from our discussion of the Fourier series I just want to mention that not all functions may have a Fourier series representation. That is to say, the Fourier series approximation may not converge on the function and in fact it may not even converge at all! I have deliberately not included the Dirichlet's theorem that defines the sufficient mathematical conditions of a function for its Fourier series representation to converge so that you can research them for yourselves. For the purposes of sanity, and the fact we study physics and not mathematics, let's assume that any function we encounter will have a convergent Fourier series. Any that don't we quickly sweep under the carpet and never speak of again (or pass them to a friendly mathematician).

7.2 Fourier Transforms

In the preceding section I mention that we can represent any general (i.e. not necessarily periodic) function as an infinite sum of its Fourier components. To do this mathematically we have to tweak the Fourier series into Fourier integrals in order for it to deal with a non-periodic function. The basic idea is that a non-periodic function can be thought of as periodic one with its period extending towards infinity, i.e. the period becomes infinitely large but not actually infinity[*]. This means that the natural frequency reduces toward zero, i.e. it becomes infinitesimally small but not actually zero. By applying this mental manipulation we can write Equation (7.7) as

$$f(t) = \sum_{n=-\infty}^{\infty} c_n e^{in\Delta\omega_0 t} \qquad (7.9)$$

where the coefficients are given by

---

[*] Strange thing, infinity ... and beyond, also.

$$c_n = \frac{\Delta\omega_0}{2\pi} \int_{-\infty}^{\infty} f(t) e^{-in\Delta\omega_0 t} dt, \tag{7.10}$$

and $\Delta\omega_0$ is our infinitesimal natural frequency. Note that I have changed the integral limits for the coefficients to reflect the idea that the period extends towards infinity and thus can also be shifted to extend to minus infinity. As the discrete values $n\Delta\omega_0$ are summed over infinity we can think of them as being mapped on to a continuous variable that, for consistency, we shall simply call ω. Due to this modification the infinite sum over n in Equation (7.9) becomes an integration over ω, on the infinite interval. Thus writing our Fourier *integral* in full gives

$$f(t) = \frac{1}{2\pi} \int_{-\infty}^{\infty} \left(\int_{-\infty}^{\infty} f(t) e^{-i\omega t} dt \right) e^{i\omega t} d\omega \tag{7.11}$$

where we have used the traditional symbol for the infinitesimal element of the integration over ω. At first glance of Equation (7.11) you may think "that was clever, we now have a complicated way of mapping a function back on to itself ... what's the point?". Well, if we now define some function of ω as

$$g(\omega) = \frac{1}{\sqrt{2\pi}} \int_{-\infty}^{\infty} f(t) e^{-i\omega t} dt \tag{7.12}$$

then Equation (7.11) becomes

$$f(t) = \frac{1}{\sqrt{2\pi}} \int_{-\infty}^{\infty} g(\omega) e^{i\omega t} d\omega. \tag{7.13}$$

Equations (7.13) and (7.12) define an integral transform and its inverse, respectively. These are commonly known as the Fourier transform and the inverse Fourier transform. The $1/\sqrt{2\pi}$ factor in both these integrals can actually chosen to be anything, so long as their product equals $1/2\pi$; the form shown is called symmetrical for obvious reasons. For convenience and shorthand the Fourier transform and its inverse can be written as

$$f(t) = \mathcal{F}\{g(\omega)\} \tag{7.14}$$

and

$$g(\omega) = \mathcal{F}^{-1}\{f(t)\}. \tag{7.15}$$

Note that instead we could have started with the time variable and transformed that into the frequency variable and, in which case, we would have to reverse our definitions (Equations (7.12) and (7.13)). So long as we are consistent with what is the transform and what is the inverse it doesn't matter what our original variable was to begin with.

The choice of our variables, time and frequency, in deriving these transforms was for instructive purposes only and Fourier transforms need not be restricted to them. They can be applied to other types of variables including those described by vectors (e.g. three dimensional space). For instance, if we considered the variable λ that represents the wavelength of some quantum particle in one or more dimensions then its Fourier transform would be the wave-number (or vector) k. This has important applications in solid state physics where the use of k-space or momentum-space is beneficial in understanding a number of electronic and optical properties of matter. As an aside, the reason why it's called momentum-space is due to De' Broglie (pronounced like Troy); hk, where h is Planck's constant, gives the momentum of a quantum particle.

A Fourier transform pair have several significant properties not least among them that the operation is linear. That is to say, if $f_1(t)$ has a Fourier transform $g_1(\omega)$, and similarly $f_2(t)$ has a transform $g_2(\omega)$, then the Fourier transform of $f_1(t) + f_2(t)$ is simply $g_1(\omega) + g_2(\omega)$.

Another property is the scaling relation that has an interesting physical interpretation. It can be shown that

$$\mathcal{F}\{f(\alpha t)\} = \frac{1}{|\alpha|} g\left(\frac{\omega}{\alpha}\right) \tag{7.16}$$

where α is a scaling factor that can be positive or negative. Equation (7.16) shows that if we squeeze the $f(t)$ along the t axis, i.e. $|\alpha| > 1$, then its corresponding Fourier transform broadens along the ω axis and also reduces in height by a factor of $|\alpha|$. Conversely, if $|\alpha| < 1$

then we broaden $f(t)$ and squeeze $g(\omega)$ this time increasing its height. In other words, the more localised the function is in time, say, the more delocalised it is in frequency, and vice versa. Remember that we are not restricted to time and frequency variables, we could just as correctly use 3-D spatial variables and momentum-space variables as a Fourier transform pair. In this case the more accurately we know a particles position, say, the less accurately we know its momentum. If you haven't come across a chap called Heisenberg yet and his uncertainty principle then you soon will[*]. As extra credit, what's the significance of $\alpha = -1$ if we stick with the time-frequency pair?

Other properties exist for shifting relations (moving the coordinate system) and the particular symmetries (odd or even functions) and complexities (real and/or imaginary functions) that I leave to the reader to investigate.

7.3 The Discrete Fourier Transform

As with all numerical procedures we first have to find a way of representing a continuous variable as a discrete set of points. Note that in doing we will not be computing the true Fourier transform but we intend to find a reasonable approximation to the transform.

Let's consider $f(t)$ as a time-dependent physical quantity obtained from actual measurements such that we have N data points taken at equidistant increments of Δt. In other words, we have the data points $(\Delta t, f(m\Delta t))$, $m = 0,1,2\ldots N-1$. Assuming that we have sufficient data points to adequately describe the behaviour of the function over a given length of time T, and that the function is periodic beyond this region, we may then use the notion that

$$\Delta\omega = \frac{2\pi}{T} = \frac{2\pi}{N\Delta t}. \tag{7.17}$$

---

[*] $\Delta x \Delta p \geq h/2\pi \equiv \hbar$

Under these conditions we can write the Discrete Fourier transform (DFT) and its inverse as

$$f(m\Delta t) = \frac{1}{\sqrt{N}} \sum_{n=0}^{N-1} g(n\Delta\omega) e^{i 2\pi mn/N} \qquad (7.18)$$

and

$$g(n\Delta\omega) = \frac{1}{\sqrt{N}} \sum_{m=0}^{N-1} f(m\Delta t) e^{-i 2\pi mn/N} \qquad (7.19)$$

where we have kept the symmetric form; in this case the product of the factors has to equal $1/N$. For convenient notation let us now drop the Δt and $\Delta\omega$ in the function arguments and use the corresponding integer multiple as a subscript instead, i.e. $f(m\Delta t) \rightarrow f_m$ and $g(n\Delta w) \rightarrow g_n$.

To implement Equations (7.18) and (7.19) into a computer program it is convenient (but not necessary) to separate the functions into their real and imaginary parts. This makes the coding somewhat more intuitive and means that we only have to deal with real numbers (imaginary numbers are essentially a real number multiplied by $i = \sqrt{-1}$, which we can drop in a computer program). In separating the real and imaginary parts we obtain the following:

$$\text{Re}(f_m) = \frac{1}{\sqrt{N}} \sum_{n=0}^{N-1} [\text{Re}(g_n)\cos(\theta) - \text{Im}(g_n)\sin(\theta)]; \qquad (7.20)$$

$$\text{Im}(f_m) = \frac{1}{\sqrt{N}} \sum_{n=0}^{N-1} [\text{Im}(g_n)\cos(\theta) + \text{Re}(g_n)\sin(\theta)]; \qquad (7.21)$$

$$\text{Re}(g_n) = \frac{1}{\sqrt{N}} \sum_{m=0}^{N-1} [\text{Re}(f_m)\cos(\theta) + \text{Im}(f_m)\sin(\theta)]; \qquad (7.22)$$

and

$$\text{Im}(g_n) = \frac{1}{\sqrt{N}} \sum_{m=0}^{N-1} [\text{Im}(f_m)\cos(\theta) - \text{Re}(f_m)\sin(\theta)] \qquad (7.23)$$

where $\theta = 2\pi nm/N$.

The code contained in the file *dft_prog.f90* performs the DFT (in one dimension) on the (normal) Gaussian distribution function with the parameters specified, then performs the inverse transform to check the correctness of the programming. The output is somewhat uninteresting in the sense that we have mapped the function back on to itself but at least it shows we've coded the DFT correctly. Note that in our implementation we have condensed the factors $1/\sqrt{N}$ into a single factor of $1/N$, which can either multiply the transform or the inverse but not both.

The DFT though straight forward to program is not terribly efficient in terms of computational effort. Each individual component of the transform requires that we sum over the N data points of the signal, and there are N such components. This leads to an operation count that is proportional to N^2; you can see this in the subroutine code where we have the nested 'for' loops. This situation only gets worse as you increase the number of dimensions in your data. How do we get around this limitation ...?

7.4 The Fast Fourier Transform

7.4.1 Brief History and Development

The fast Fourier transform (FFT) has been independently discovered and rediscovered by various people, the most earliest version appearing in the literature being attributed to Gauss in 1866[*]. For whatever reasons Gauss's idea was largely ignored by the scientific community and no-one connected it to the use of modern computation. In 1965 the American mathematicians James William Cooley and John Wilder Tukey published an article that discussed in

---

[*] It appeared as an unpublished manuscript in his collected works. The actual date Gauss wrote this manuscript is presumed to be around 1805, which actually *predates* Fourier's original work by two years – something to think about.

details the use of a machine algorithm to calculate complex Fourier series. This is largely credited as the first formal use of the FFT on a "modern" computer. However, more than twenty years before a pair of physicists (ok, ok and mathematicians) Cornelius Lanczos and Gordon C. Danielson* gave a particularly lucid description of the FFT derivation in their 1942 publication on practical Fourier analysis of X-rays scattered from liquids. It goes a little something like this...

Let's assume N is an even number. We can then write the DFT as a summation over the even-numbered points and a summation over the odd-numbered points. Mathematically this is written as

$$g_n = \sum_{m=0}^{(N/2)-1} f_{2m} e^{-i2\pi n(2m)/N} + \sum_{m=0}^{(N/2)-1} f_{2m+1} e^{-i2\pi n(2m+1)/N}$$

$$= g_n^{(even)} + g_n^{(odd)} e^{-i2\pi n/N} \qquad (7.24)$$

where we define

$$g_n^{(even)} = \sum_{m=0}^{(N/2)-1} f_{2m} e^{-i2\pi nm/(N/2)} \qquad (7.25)$$

as the even-numbered points, and

$$g_n^{(odd)} = \sum_{m=0}^{(N/2)-1} f_{2m+1} e^{-i2\pi nm/(N/2)} \qquad (7.26)$$

as the odd-numbered points. Note that we've ignored the $1/\sqrt{N}$ factor here, which can be easily reintroduced at a later stage. Let's take stock of what we have just done. By splitting the DFT into even and odd summations we have essentially produced two new DFTs, Equations (7.25) and (7.26), with half the number of points of the original transform. Hence the number of operations required is now proportional to $2 \times (N/2)^2$ i.e. half the original. The beauty of this algorithm is that we can keep going and further split those new DFTs into their even and odd-numbered points, and so on until we reach the level where there is only one component to find in the summation. However, this requires that the number of points at

---

* Nothing to do with Mr. Miyagi or the Cobra Kai – ask your parents

Fourier Analysis

each subdivided level contained within the summation remains even. This can easily be insured by specifying that N is an integer power of two. For instance let $N = 2^k$ then after k subdivisions there will be N DFTs to compute each with only one component to find. In other words instead of the operation count being proportional to N^2 it is now proportional to Nk or more generally $N \log_2 N$. The FFT is indeed fast!

7.4.2 Implementation and Sampling

The reason why Cooley and Tukey are generally credited with the discovery of the FFT as applied to modern computing was their clever way of interweaving the summation pairs at the lowest level of the algorithm. This interweaving is really just an exercise in bookkeeping which is rather tedious, and can make the coding somewhat complicated. Rather than discussing the interweaving strategy at length I have provide the program file *fft_mod.f90*, which contains a FFT subroutine, for your use. For interested readers may I point you in the direction of Landau's book pages 256 to 263 for an in-depth discussion of the interweaving strategy.

To check that the FFT subroutine works correctly write a program that passes to the subroutine an array with only one non-zero value located in the first half of the array, except the first entry. The array will need to be of type double complex (use the code provided as a guide) and of a size defined by an integer power of two; this integer is also passed to the subroutine as an argument. The integer flag INV needs to be set to one and you will have to save the data to file in order to visualise it elsewhere. Be aware that a complex data type array, though declared as a single dimension, is actually a two column array. The first column contains the real number and the second column contains the imaginary number[*]. To output the Fourier spectrum using Fortran we simply print out or write to file the absolute value (ABS) of the complex array. Additionally, you will have to map the integer indices of the outputted array on to the inverse Fourier transformed variable; as a hint think of the input

---

[*] More precisely their real coefficients – we can't store the $\sqrt{-1}$ on a computer!

array indices as the discrete frequency spectrum of a time varying signal.

So did you get what you expected? You should have found that the larger the size of your input array the more resolved the time signal for a particular (discrete) frequency. This is actually a consequence of Equation (7.17) whereby our $\Delta\omega$ is a constant and equal to one (the input array indices) thus increasing N implies that Δt decreases.

Now that we have established the inverse FFT works for a frequency spectrum into a time signal let us try the other way around. However, before we continue I just want to briefly describe how to interpret the spectrum resulting from the FFT subroutine. It assumes that the (time) data passed to it is periodic on the interval for which it is defined. Thus the outputted spectrum is also periodic. Figure 7.2 illustrates this point. Here we imagine a sketch of the frequency spectrum of some arbitrary harmonic (time) signal with natural frequency ω_0 calculated using the FFT subroutine*. The black curve on the positive frequency portion of the plot is the complete array output from the subroutine. Note that we have mapped the output array indices on to (discrete) frequency values including zero, meaning that the total size of the array spans up to $(N-1)\Delta\omega$ on the frequency axis. As the spectrum is periodic all the data after $(N/2)\Delta\omega$ is the same as the spectrum at *negative* frequencies, shown in the shaded region of Figure 7.2. As a consequence you can think of the subroutine as computing both the normal, forward time and time *reversed* frequencies; essentially a mathematical quirk of the FFT algorithm. Remember that we count the zero frequency as a positive frequency. To demonstrate we can write

$$\cos(\omega_0 t) = 0.5\cos(\omega_0 t) + 0.5\cos(-\omega_0 t) \tag{7.27}$$

as cosine is an even function. Conversely

$$\sin(\omega_0 t) = 0.5\sin(\omega_0 t) - 0.5\sin(-\omega_0 t) \tag{7.28}$$

---

* The broadening of the peak is due to a phenomenon termed leakage that we will discuss shortly.

as sine is an odd function. As the spectrum is shared between the positive and negative frequencies it's intensity (i.e. it's Fourier coefficient value) is half what we would expect if we just consider the "physically" significant positive frequencies.

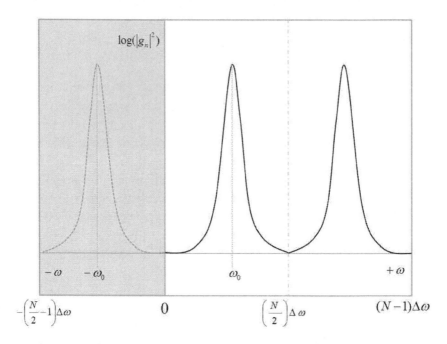

Figure 7.2: Sketch of a frequency spectrum of some harmonic oscillator with natural frequency ω_0. We have deliberately included peak broadening to clearly demonstrate the interpretation of the spectrum.

The upshot of all this is that when recording the spectrum data we should only store the first $N/2$ points and multiply their values by two in-order to obtain the "physically" correct power spectrum.

Modify the program you have written to analyse the spectrum of the following function

$$f(t) = \cos(5\pi t), \qquad (7.29)$$

sampled once per second for 32 seconds, then twice per second for 16 seconds, and so on up to 16 times a second for 2 seconds. Here we keep N constant at 32. As we are taking a transform rather than the inverse you will have to change the INV flag to zero. Also we are now mapping the input array indices, which represent the time sampling, to discrete frequencies you will have to calculate these accordingly. To properly view the spectrum we should plot the square of the magnitude of the transform against the *discrete* frequencies. In other words, the spectrum can be considered a histogram with the width of each (frequency) bin given by $\Delta\omega$ and its height given by

$$|g_n|^2 = \text{Re}(g_n)^2 + \text{Im}(g_n)^2. \tag{7.30}$$

At higher frequency resolutions, i.e. narrow bin widths, you may find it more intuitive to plot the spectrum as a line graph mimicking the idea that we are computing an approximation to a continuous variable. It may also be advantageous to plot the spectrum on a logarithmic axis.

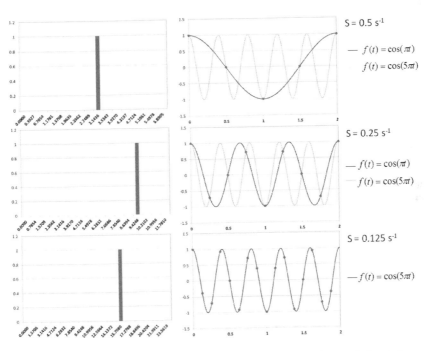

Figure 7.3: Left - the frequency spectrums of Equation (7.29) sampled at twice, four times, and eight times per second respectively. Right - the time sampled function with the detected frequency harmonic shown.

140

After performing the FFT on the given function you should obtain results similar to Figure 7.3. We know that the (angular) frequency of the function described by Equation (7.27) has to be $\omega = 5\pi \approx 15.7$. Why then do we see a frequency of π in the spectrum when sampling at a rate of twice a second, and indeed a frequency of 3π when sampling at four times a second? The answer lies in the plot of the time sampled function overlaid on the actual function as shown in the right hand column of Figure 7.23 When sampling at twice per second (top) we see that the sampled data resemble a triangular waveform with a time period of 2 seconds, equivalent to a harmonic frequency of π. Similarly, when sampling at 0.25 per second (middle) the curve $f(t) = \cos(\pi t)$ pass through the sampled points leading to the erroneous frequency spectrum. This is called *aliasing*; the higher frequency "signal" has been aliased by lower frequency harmonics. This happens because, in these two cases, we are *undersampling* the function; our sample rate is not sufficiently high to capture the true waveform. As a rule you should sample the signal at a rate at least twice the highest frequency component contain in the signal. In our example the angular frequency of our waveform is 5π equivalent to a frequency of 2.5 Hz. Thus to avoid undersampling we should take data at time intervals at least 0.2 seconds apart or less. Indeed, when we sample at intervals of 0.125 seconds we recover the correct frequency spectrum.

What then happens when we sample at a rate of 32 times per second for one second? The FFT has filled all the discrete frequency bins equally. This problem is known as *leakage* and occurs when there is a lack of frequency resolution such that actual frequency of the data does not match one of the frequency bins. In this case the FFT tries to compensate by distributing the transform across nearby frequencies, in other words, it leaks. To alleviate this problem we can increase the total observation time, that is we increase N but without changing the sampling rate. Try sampling at the same rate of 32 per second but for two seconds, i.e. increase N to 64, and see if we get a better outcome.

The leakage problem can be attributed to where in the time domain we finish sampling. Remember that the FFT assumes the data you pass to it is periodic on the observation interval for which it is defined. If the sampling finishes mid-period then the FFT "sees" a discontinuity in the function. As we know from the Fourier series a discontinuity is better approximated by increasing the number of terms in the sequence. Comparatively, the FFT increases the number of frequencies detected in the spectrum to deal with the discontinuity. It is therefore advantageous to use a sampling rate that is proportionate with the period of the function.

Exercises

1. Calculate the Fourier series for a Saw-tooth waveform. Plot the results to get a clear picture of how the series converges to the function.

2. Investigate the overshoot in the Square wave form as the number of terms in the Fourier series increases. Calculate the error between the series approximation and the function, and hence determine the behaviour of the overshoot as we retain more terms in the series – this is called the Gibbs phenomenon.

3. Derive the Fourier coefficients of Equations (7.2) and (7.3). Hint: sine and cosine functions are orthogonal.

4. Derive the scaling property for the Fourier transform; Equation (7.16). Then derive the similar property for the inverse transform.

5. Find a way to time the operation of a program in Fortran then evaluate the runtimes of the DFT algorithm versus the FFT algorithm for the same set of data. Check the statements that the DFT algorithm operation count is proportional to N^2 and the FFT operation count is proportional to $N \log_2 N$. Also check that the output from each algorithm is the same for the same input (within unit round-off error of the computer).

6. Use the FFT subroutine to obtain the spectrum of the function $f(t) = \sin(5t)$. Use a sampling rate that is sufficiently rapid to avoid undersampling. Can you derive a sampling rate that avoids the problem of leakage?

7. Consider the function $f(t) = \cos[(1 + \alpha)t] + 2\cos[(2 - \alpha)t]$, for α in the range $[0,1]$. Investigate how the sampling rate and overall observation time affects the resolution of the frequency peaks.

8 Monte Carlo Methods

"Throw enough s**t at the wall and some of it's bound to stick."
                                                    - Anonymous

Monte Carlo* methods (or Monte Carlo experiments) are a broad class of computational algorithms that rely on repeated random sampling to obtain a numerical result. They are often used in physical and mathematical problems when it is impossible to obtain an analytical solution, and application of a direct algorithm is infeasible. Monte Carlo methods are mainly used in three distinct problems: numerical integration, simulation, and optimisation. Here we discuss the first two in this list and how they relate to physics problems.

8.1 Monte Carlo Integration

8.1.1 Dart Throwing

"Hit and miss" integration, also known as the shooting method, is the most intuitive type of Monte Carlo method to understand. To demonstrate the application of this approach lets discuss a novel way of approximating the value for π. Consider Figure 8.1. It shows the upper right quadrant of a circle of unit radius circumscribed by a unit square. Imagine throwing darts randomly at this board (some of you may have had a similar experience already in the student's union bar). Of the total number of darts that hit within the square, the fraction of those that land within the circle will be approximately equal to the ratio area of the circle contained by the square. Mathematically we write

$$\frac{A_{circle}}{A_{square}} \approx \frac{N_{circle}}{N_{thrown}}. \tag{8.1}$$

---

* Named after the famous casino for reasons that will become apparent.

Here we have the constraint that darts cannot be thrown outside of the square and A_{circle} is the area contained in the unit square.

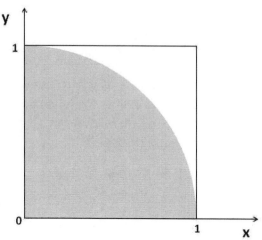

Figure 8.1: The Monte Carlo "dart board" used to approximate π.

Remembering your geometry basics we can substitute and rearrange the equation above to give a approximation formula for π, such that

$$\pi \approx \frac{4 N_{circle}}{N_{thrown}} \qquad (8.2)$$

In other words, the probability that a dart will hit the shaded area is equivalent to one quarter of the value of π. Despite the fun you can have in trying to make the dart throwing random, attempting to physically perform this experiment soon becomes tedious as you need a large number (a thousand, at least, even then convergence isn't guaranteed) of thrown darts to get a reasonably accurate approximation for π. Instead we make a computer simulate the dart throwing by having it generate random numbers.

Now before anyone gets militant on my personage computers don't actually generate random numbers since they are deterministic. Computers can generate what are known as pseudorandom numbers via a recursion formula; given a starting point, generally referred to as the random number seed, the generator produces a sequence of "random" numbers by performing mathematical operations on the previous "random" number. Rigorous statistical tests can be applied to the outputs of these generators to check that the numbers are random in relation to one another. As a cautionary note, a random number generator will produce the identical random number sequence for the same seed. Hence, for multiple trials different seeds have to be found in order to produce different random number sequences. Typically this is done by using the system's clock. Fortran has a couple of built in functions that perform random number generation that will be adequate for our purposes. As whole books have been dedicated to the topic of random number generation and how to seed them I will leave the topic there and return to the discussion of integration via Monte Carlo.

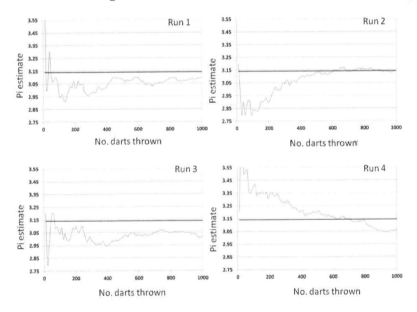

Figure 8.2: Results of estimating from a Monte Carlo integration for four separate runs. The black line represents π.

For each random throw we generate two random numbers, x and y, that represent the displacement from the origin to where the dart hit in the horizontal and vertical directions, respectively. Using Pythagorean Theorem we can calculate the distance from the origin and thus determine whether or not the dart landed within the circle. That is, if the distance is greater than one unit it missed, less than or equal to one unit it hit. By keeping count of the total number of darts thrown (i.e. the random (x, y) coordinate generated) and the number that hit the circle we can approximate π using Equation (8.2).

The file *PI_Monte.f90* contains a program to perform this experiment. The code generates a pair of (uniformly distributed) random numbers both on the interval [0,1], to simulate where the thrown dart lands within the unit square. After computing the distance from the origin we either add one to the counter if it's a hit or do nothing if it's a miss. After every 10 darts are thrown we estimate π using Equation (8.2) and store the results to file. Note also that we can save the random numbers generated to check the randomness of the throws. In this file you will also find a subroutine to generate a random seed number based on the system clock. Feel free to use this in any of your own code. Note the function 'SECNDS(0.0)' returns the time in seconds since midnight of the *current* day. Thus should you just so happen to run the program at the exact same second since midnight on separate days you would generate the same sequence of random numbers.

Figure 8.2 shows the results from running this program on four separate occasions, using a total of 1000 dart throws. Here we can see that our random number seeding and generation has done an adequate job; the four different runs have produced four *different* results as we should expect if we had physically performed the experiment on four separate occasions.

The black line in each of these plots represents the actual value of π. The figure suggests that although we are not guaranteed to converge on the actual value of the integration, the approximation does, to some extent, stabilise as we increase the number of throws. However, remember that the results are *accumulated*. Thus, as the number of throws increases the influence of the next throw is reduced, and the variation in the estimate from one throw to the next necessarily diminishes. In other words, the results at the end of the experiment is very much influenced by the outcome of the throws at the start of the experiment.

If you're looking at the results plotted in Figure 8.2 and wondering why we would bother with Monte Carlo integration at all remember that this is an illustrative (and simple) example. If the integration can be done easily by other means then the Monte Carlo method should *not* be used. The Monte Carlo integration comes into its own when other numerical techniques are difficult, if not impossible to implement. That said we can still use our simple example to discuss how to estimate the accuracy of the Monte Carlo integration.

Figure 8.3 plots histograms of performing 10,000 dart throwing integrations with 100 darts per integration in the top panel and 1000 darts per integration in the bottom panel. Note that both plots are over the same range, but they have different bin widths; 0.1 for the top plot, and 0.01 for the bottom plot, which are related to the number of darts thrown per integration.

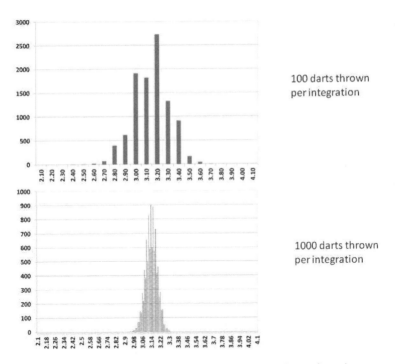

Figure 8.3: Histograms of 10,000 Monte Carlo integrations of π using 100 darts per integration (top) and 1000 darts per integration (bottom).

Monte Carlo Methods

You should recognise the curves as being the bell shape of a normal (or Gaussian) distribution; in fact with an increasing number of integrations the distributions would become much smoother and would approach the ideal bell shape. As can be seen from the plots as we increase the number of darts thrown per integration the distribution of estimates for π narrows about the true value. In other words the mean value of the distribution becomes a more accurate value for π. The question then is how much more accurate...

From Figure (8.3) you can convince yourself that the width of the distribution at half height for the 1000 darts per integration case is roughly 1/3 of that for the case of 100 darts per integration. If you remember your probability theory you should recall that the width of a normal distribution of estimates of a value is proportional to one over the square root of the total number of points used to compute each estimate. In other words, the factor difference between the widths of these two distributions should be equal to $1/\sqrt{10}$, which is what we find.

Moreover, the standard deviation of the mean, which is a measure of the width of the distribution, can be itself estimated from a *single integration* using

$$\sigma_N = \sqrt{\frac{\frac{1}{N}\sum f_i^2 - \left(\frac{1}{N}\sum f_i\right)^2}{N-1}} \qquad (8.3)$$

where f_i is the estimate of the value (π in our case), after the ith point is sampled (dart is thrown), and N is the number of points sampled (darts thrown) in total. For very large N we can drop the minus one in the denominator. It is of note that Equation (8.3) can be updated after each new random point is sampled; in which case N becomes equal to the value of i we have reached. This means we can monitor the confidence we have in the estimate of the integrated value as we increase N. Remember that the estimate lies within σ of the precise average to a 68.3% degree of confidence; within 2σ to a 95.4% degree of confidence; within 3σ to a 99.7% degree of confidence; and so on. To decrease σ and therefore improve the accuracy in the estimate we merely have to sample more random points. The drawback to this method is that the improvement can only go as the square root of N. This takes us back to the point made previously that there is a law of diminishing returns due to the accumulation of data; when we've already sampled 1000 points, say, one more sampled point makes little difference to the outcome, but another 1000 would.

8.1.2 General Integration using Monte Carlo

In our shooting method example above to estimate π we have, in a round-about fashion, approximated the integral

$$\int_a^b f(x)dx = \int_0^1 \sqrt{(1-x^2)}dx = \frac{\pi}{4}. \tag{8.4}$$

A slightly more direct method of using the Monte Carlo integration would be to sample (uniformly distributed) random values of x on the interval $[a, b]$, and finding the average of the function evaluations, $f(x)$. In general, for a one dimensional integration, we are using the notion that

$$\int_a^b f(x)dx = (b-a)\langle f \rangle \tag{8.5}$$

where $\langle f \rangle$ is the precise average of the function on the interval $[a, b]$. This has a very straightforward geometrical interpretation as depicted in Figure 8.4.

The integration, which is the area under function defined on the interval, is equal to the area of the shaded rectangle. The Monte Carlo method is an attempt to estimate the precise function average, $\langle f \rangle$. Notice that as a consequence of this interpretation $\pi/4$ must be the precise function average of the unit, quarter circle.

Formally we write the Monte Carlo integration estimate as

$$\int_a^b f(x)dx \approx \frac{(b-a)}{N}\sum_{i=1}^{N} f(x_i) \qquad (8.6)$$

where N is the total number of randomly sampled points.

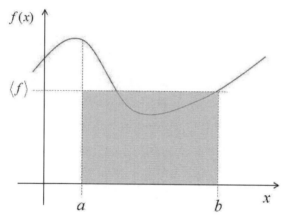

Figure 8.4: Geometrical interpretation of Monte Carlo integration.

In the file *Monte_Carlo.f90* you will find a module that contains a subroutine to perform the Monte Carlo integration according to Equation (8.6). You can check that it works by choosing an easily analytical integral and seeing if we obtain the same result using the Monte Carlo method. We could also use this information to check how well the estimate for the standard deviation models the actual error in the integration.

The Monte Carlo method of integration is most effectively used in the computation of multidimensional integrations where the application of more direct methods is either infeasible or impossible. To perform a multidimensional integration via the Monte Carlo method we simply find random numbers for all the variables involved, find the value of the function at those coordinates, then update the sum. For instance a two dimensional integration can be written as

$$\int_c^d \int_a^b f(x,y) dx dy = \frac{(d-c)(b-a)}{N} \sum_{i=1}^{N} f(x_i, y_i). \qquad (8.7)$$

As we add more dimensions we generate more random numbers for the additional dimensions and multiply by the relevant integration interval; the sum over f divided by N still proves an estimate of the precise function average defined with the integration region. Note that because of this simplicity the Monte Carlo method of integration has some inherent advantages over more direct numerical techniques ...

The error in the direct methods for numerical integration (trapezoidal rule, Simpson's rule, etc.) stems from the number of terms retained in the Taylor series approximation of the integrand function. For example, the trapezoidal rule approximates the integrand with a linear polynomial, which, as we have seen, has an error that is $\mathcal{O}(h^2)$, where h is the strip width across the integration interval. If we wish to halve the error in our numerical approximation of the integration in one dimension we decrease h by a factor of $\sqrt{2}$; this is equivalent to increasing N (the number of function evaluations) by the same factor. For a two dimensional integration to halve the error we have to apply this modification in both dimensions such that N increases by a factor of 2 in total. For three dimensions N has to be increased by a factor of $2^{3/2}$. In general, for a d dimensional numerical integration to halve the error in our approximation we would have increase N by a factor of $2^{d/m}$, where m represents the order accuracy of the numerical integration method used to compute the approximation (for the trapezoidal rule $m = 2$; Simpson's rule $m = 4$; and so on).

The error in the Monte Carlo method is different. As we have just discussed the error produced by a Monte Carlo computation is probabilistic in nature; we can say that the approximation calculated is within one standard deviation of the "true" value 68.7% of the time. To improve the approximation, that is to reduce the standard deviation and thus make the average value converge on the "true" value, we increase the number of random points sampled, N. As this is a probabilistic process we know that the error will reduce as $\sqrt{N}$. Thus, to halve the error we increase N by a factor of 4. *This is independent of the dimensionality of the integration*! To explain, we perform a multidimensional integration via the Monte Carlo method by computing as many random numbers as there are dimensions, then evaluating the function at the coordinates specified by those random numbers and updating the sum. Note that this is a true scatter shot approach; none of the dimension variables are held constant, we just keep "shooting" and evaluating a single value for the function at those coordinates randomly generated.

To demonstrate, let's imagine a four dimensional integration that we can perform either by the trapezoidal rule or the Monte Carlo method. After obtaining the approximation from both methods we'd like to halve the error in each. From our discussions we can see that both require the number of function evaluation points to be increased by a factor of 4. In other words the rate of convergence for the two methods is comparable when performed on a four dimensional integration. To avoid confusion, here we are talking about the rate of convergence of an approximation to the "true" value, not the absolute value of the error. It is likely that the trapezoidal rule is more accurate (has less absolute error) than the Monte Carlo method to start with. That said, for dimensions higher than four the Monte Carlo method will actually converge more rapidly than the trapezoidal rule. Indeed, for integrals of sufficiently high dimensionality the Monte Carlo method will converge more rapidly than any direct method we have discussed, dependent on the method's order of accuracy.

A secondary advantage to the Monte Carlo method is the number of function evaluations that have to be performed in order to gain an approximation to the integral. To illustrate, imagine a 10 dimensional integration that we are computing via the Monte Carlo method. Let's say we evaluate the integrand function 10 times, that is, we generate 10 random numbers for each function evaluation i.e. 100 points in total. Now we want to halve the error in our approximation so as stated we increase N by a factor of 4, i.e. we have to generate 400 random numbers. If we evaluated the same integration using the composite trapezoidal rule, again with 10 function evaluations per dimension, then would have $10^{10} = 10$ billion function evaluations to perform. To halve the error we would have to increase N by a factor of 2^5; we would now have to perform 320 billion function evaluations! Now while it's true to say that the trapezoidal rule will, in all likelihood, give a more accurate result than the Monte Carlo method it would certainly take more time (an infeasible amount) to obtain. Even though the Monte Carlo estimate will be crude, the method does give a quantifiable measure of the error, and the knowledge that we can improve this error by taking just a few more (certainly less than billions of) randomly sampled points.

8.1.3 Importance Sampling

Before we leave Monte Carlo integration I would just like to mention a technique that can help improve the accuracy of the method called importance sampling. Importance sampling using information about the function to place more randomly sampled points where the function is largest, meaning that the approximation is more accurate for the same number of sampling points. To do so we find an function g(x) that approximates the integrand function f(x) over the integration interval so that we can write

$$\int_a^b f(x)dx = \int_a^b \frac{f(x)}{g(x)} g(x)dx = \int_{y^{-1}(a)}^{y^{-1}(b)} \frac{f(y^{-1})}{g(y^{-1})} dy \qquad (8.8)$$

where

$$y = \int^x g(t)dt . \qquad (8.9)$$

Interpreting Equation (8.8) we see that instead of integrating $f(x)$ over x, we integrate the ratio $f(x)/g(x)$ over y. This has the geometrical effect of flattening the integrand over the integration interval, thus making the Monte Carlo method more accurate. Remember we are attempting to approximate the precise function average over the integration interval. Having a "flat" function makes this easier.

For example, consider the integral

$$I = \int_0^{\pi/2} \sin(x)\,dx. \tag{8.10}$$

This has the analytical value of 1. We can approximate sine using the first term of its Taylor series expansion $\sin(x) \approx x$, such that the integral becomes

$$\int_0^{\pi} \sin(x)\,dx = \int_0^{\pi^2/8} \frac{\sin(\sqrt{2y})}{\sqrt{2y}}\,dy, \tag{8.11}$$

where

$$y = \int_0^x t\,dt = \frac{x^2}{2} \tag{8.12}$$

and

$$x = \sqrt{2y}. \tag{8.13}$$

Without importance sampling, the Monte Carlo method using 100 sampled points, produces a result with $|3\sigma| \approx 0.15$. For the same number of sampled points but with importance sampling we obtain a result with $|3\sigma| \approx 0.04$. Note that the approximation function $g(x)$ should be a reasonably good across the entire integration interval otherwise the result is likely to be worse (see Exercise 1).

8.2 Monte Carlo Simulations

One of the first uses of a Monte Carlo method was in determining the thickness of the shielding required to stop neutrons from leaking from a nuclear reactor. Developed in the 1940s by Stanislaw Ulam when he worked on the Manhattan project it coincided with the birth of modern computing. Indeed, Jon Von Neumann was the first to successfully program a Monte Carlo method on ENIAC (Electronic Numerical Integrator and Computer) in the late 1940s and into the 1950s.

The process of a neutron travelling through metal is very much a random one; the neutron collides with the metal atoms and is scattered in a random direction. Any influence of the neutron's previous motion is lost in the (random) scattering process i.e. there is no correlation between the results of a particular collision and the neutron's initial motion. This kind of random process is called stochastic and they frequently occur in physics; molecular diffusion; percolation of atoms on (growth) surfaces; and radioactive decay, to name but a few.

8.2.1 Random Walk

Let's begin our discussion of Monte Carlo simulations with a simple drunken walk. As some of you may have already found out, a drunken walk can be described as random experience. Mathematically speaking a random walk is one in which you are equally likely to step in any direction. The question normally posed is how far you will travel in a given number of steps. (Though the more pertinent questions might be how long will it take you to get home, and will you have sobered up by then?)

To answer this question we simulate the walk using random numbers, and, of course, some assumptions and constraints. The first assumption is that each of your strides is equal in length, and that you only walk along the cardinal directions. In other words, you are on a two dimensional unit square grid moving from point to point. After reaching each point you have equal probability of going north, south, east, or west. This can be simulated by generating a random number on the interval [0,1], with equal intervals defining the direction taken e.g. 0 to 0.25 walk north, 0.25 to 0.5 walk east, and so on. As the simulation runs we keep track of the x and y distances travelled from our starting point (origin) and calculate the distance travelled using Pythagoras, either at the end of the run or updated after each step. As this is a stochastic process we don't gain much insight from one random walk, and so we should run the simulation many times over to obtain statistically valid results.

After writing such a program we obtain the results presented in Figure 8.5. Here we have results for $N = 10$ up to $N = 1000$ in increments of 10 where each value plotted is the average of 1000 simulations. Although this drunken walk scenario may seem somewhat over simplified can you think of any real physical situations to where it might be applied? (Hint: the grid need not be a square grid but a regular grid of some other shape that may have more possible directions of travel.)

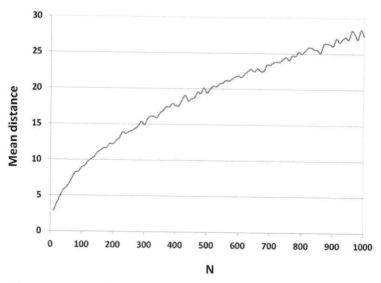

Figure 8.5: Mean distance travelled versus number of steps taken in a drunken walk

We can relax the constraint that we only walk along the cardinal directions and allow any direction on the two dimensional surface, maintaining a stride length of one. To do this we consider the angle that governs the direction of our next step. In other words, we randomly sample the angle φ on the interval $[0,2\pi]$, and use trigonometry to determine the x and y distances moved per step i.e. $x = \cos(\varphi)$ and $y = \sin(\varphi)$. Figure 8.6 shows the effect of the modification to the results of the mean distance travelled against the number of steps taken under the same conditions as shown in Figure 8.5. Notice something familiar?

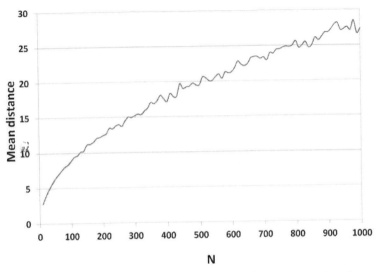

Figure 8.6: Mean distance travelled versus number of steps taken for a random walk on a 2d surface where any direction is possible.

We can also remove the assumption that the stride length is a constant by instead of randomly sampling the angle of travel, we randomly sample two numbers per step on the interval $[-1,1]$ that represent the x and y components of a displacement vector. Note that we could also do this by uniformly sampling theta on $[0,2\pi]$ and uniformly sampling the stride length, r, on the interval $[0,1]$. However, these two methods are subtly different; can you spot why, and does this affect our results in any way?

What your results should have told you is that the qualitative result is unaffected by the size of the strides. In other words, we have scale invariance and we can simply set stride length to unity. The total mean distance travelled is then measured in units of the stride length. In technical parlance, our stride length is what is known as the mean free path.

Now that we have covered a random walk over a two dimensional surface let's try to extend that to a random walk in a three dimensional volume. In this case, it is useful to think of a perfume molecule diffusing through air; the randomness, or stochastic process, is introduced by the perfume molecule randomly scattering from collisions with the molecules that make up the air. If we assume that the perfume molecule is equally likely to scatter in any direction then we have spherical symmetry.

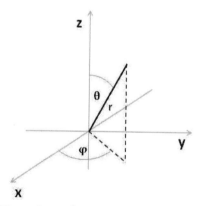

Figure 8.7: Spherical coordinate system

Figure 8.7 shows the spherical coordinate system (r, θ, φ), where r is the length of some vector, θ is the polar or inclination angle (angle between the z axis and the vector), and φ is the azimuth angle (angle between the x axis and the projection of the vector on to the x-y plane). As we have discussed the mean free path can be set to a constant that is equal to one, i.e. $|r| = 1$. We therefore need to uniformly sample random numbers on the surface of the unit sphere.

If we naively sampled θ from $[0, \pi]$ and sampled φ over $[0, 2\pi]$ both with uniform distributions we would run into problems. The issue stems from the fact that the surface of a sphere is curved (you can't flatten it on to a 2D plane) and can explained as follows. In our two dimensional representation of the problem the azimuth angle, φ, can be sampled uniformly on the interval $[0, 2\pi]$ as this leads to a uniform distribution of points on the circumference of the (unit) circle (see Figure 8.8(a)). Now consider Figure 8.8(b). This shows a sphere cut by the *x-y* plane such that a circle can be drawn around the sphere, which defines its equator. If we now lift the *x-y* plane up through the sphere, as if decreasing the polar angle θ by uniform increments, the circle defining where the *x-y* plane cuts the sphere necessarily contracts. If there were a uniform distribution of points on the circle then as the circle moves up the sphere those points also contract. In other words, the distribution of points on the *surface* of the sphere would be not uniform and in fact would bunch at the poles.

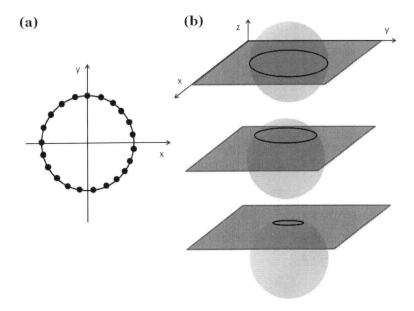

Figure 8.8: (a) Points on the unit circle defined by a random sampling of φ on the uniform distribution $[0, 2\pi]$. (b) Unit sphere cut by the *x-y* plane; the circle defining the intersection deminishes as the plane is pulled upwards.

The way around this problem is not to randomly sample numbers for θ on a uniform distribution but to sample points on a $\sin(\theta)$ distribution. We can see this from the figure; we need less points at the poles where $\theta = 0$ and π, and more points around the equator where $\theta = \pi/2$.

More formally, we consider the solid angle element

$$d\Omega = \sin(\theta)d\theta d\varphi. \tag{8.14}$$

We would like to uniformly sample on the element $d\Omega$, which means that we need to uniformly sample both $\sin(\theta)d\theta$ and $d\varphi$. If we let

$$dg = \sin(\theta)d\theta, \tag{8.15}$$

then we can write

$$g(\theta) = \cos(\theta). \tag{8.16}$$

We now have the means to uniformly sample of the surface of the (unit) sphere. We select φ from a uniform distribution on the interval $[0,2\pi]$ and select g from a uniform distribution on the interval $[-1,1]$, where θ is then calculated using

$$\theta = \cos^{-1}(g). \tag{8.17}$$

Modify your program* so that it can perform this three dimensional simulation of a molecule diffusing through the air; you will have to calculate the x, y, and z coordinates of the displacement vector using trigonometry. Again, we should be able to show a relationship between the mean distance travelled and the number of collisions taken, so long as these results are statistically valid. Once you have the plot it should be the same as those we have previously presented. How might we show that the mean distance travelled is a power of N, and how might we discern that power?

These data *suggest* that we have stumbled upon a principal property of nature, that the average distance travelled of a particle undergoing

---

* If you haven't written a random walk program yet, do so.

random scattering events is proportional to the square root of the total number of scattering events to which it has been subject. Mathematically, we write

$$\frac{\langle R \rangle}{\lambda} \approx \sqrt{N}. \tag{8.18}$$

Note the use of the operative word suggest. We haven't proved anything only that we have statistically valid results. What our computations have given us is significant insight into the physics of diffusion, and a strong suggestion that the relationship described by Equation (8.18) probably is real.

This outlines the importance of Monte Caro simulations; it can provide qualitatively valid results for physical systems that may be difficult or impossible to solve using more direct methods (both analytical and numerical).

That said there is more information we can extract from our simulations that lends credibility to the results we obtained. By simulating many molecular paths we have actually modelled the situation whereby a bottle of perfume has been opened and the many molecules of the aroma have diffused into the air (under the assumption that they all emerged from a singularity located at the origin). In which case, we should be able to visualise the distribution of aromatic molecules as a function of distance from the perfume bottle after a particular number of collisions. Essentially, this represents the density of aromatic molecules in the air as we move away from the perfume bottle at a particular moment in time. I leave this as a problem for you to solve, but before you actually calculate the distribution how might you expect it to look, and how might you expect it to evolve with time (number of collisions)?

8.2.2 Radioactive Decay

Radioactive decay occurs when an unstable atom (or particle) releases some form of radiation (alpha, beta, and or gamma) and decays into other particles. This is also referred to a spontaneous decay, in that it requires no external stimulation to occur. Each unstable atom has the same probability to decay in any given time period, but when this specifically happens is random. As the total number of unstable atoms decrease the number that decay in a particular time period also decreases. In the limit of the time period going to zero we can say that the rate of decay is proportional to the number of unstable atoms that still exist. Thus, when there are a large number of unstable atoms in a sample spontaneous decay is well modelled by exponential decay. Essentially, this is a continuous or large number approximation to the actual process of the discrete decay events. As the number of unstable atoms inevitably decrease this approximation begins to fail and the process becomes increasingly stochastic (subject to chance).

In the limits of $N \to \infty$ and $\Delta t \to 0$ we write

$$\frac{\Delta N(t)}{\Delta t} \to \frac{dN(t)}{dt} = -\lambda N(t), \qquad (8.19)$$

where λ is the so called decay constant. This is related to the half-life of the unstable atoms by

$$T_{1/2} = \frac{\ln(2)}{\lambda}, \qquad (8.20)$$

and λ can be described as the activity of the unstable atoms i.e. the probability that an atom decays within a given time period of Δt.

We can use Monte Carlo simulation to model when this change in behaviour occurs. More precisely, we can determine the minimum number of unstable atoms required for the large number approximation to hold true. To do this we increase time in discrete steps, and for each of these steps we count the number of decay events that occurred during that interval. By keeping track of the number of atoms left in our simulation we can quit once they have all decayed. To simulate a decay event we generate a random number, and if that number is less than λ then a decay event occurs and we drop our atom count by one.

Unless we are comparing our computational results to actual experimental data we can ignore any time scaling that may be required. For instance, if $\lambda = 0.7 \times 10^4 s^{-1}$ then we should set our time intervals to be equal to $10^{-4}s$ so that we can set $\lambda = 0.7$ in our program, and are able to use random numbers in the range [0,1]. Otherwise, we keep the value of λ as it is and scale our random numbers accordingly; $\lambda = 0.7 \times 10^4 s^{-1}$ and the random numbers are scaled to $[0, 10^4]$. In our time scale free program the increments in time can be equal to one (i.e. a single integer) and λ is chosen somewhere between zero and one.

The file *Monte_Decay.f90* contains a program that performs this simulation. The parameters in this file should be self-explanatory and you will have to include the initialise random seed subroutine from the previous program. Make sure you understand the code as I have asked to use it in one of the exercises that follow.

Exercises

1. Evaluate the integral

$$I = \int_0^\pi \sin(x)dx$$

using the Monte Carlo method with importance sampling, where $g(x) = x$. Compare the results to the same integration without this importance sampling. What went wrong? Retain an additional term from the Taylor series expansion of sine and try again. Rather than retaining more terms in the Taylor series is there any other way to approximate sine over this interval?

2. The true period of a pendulum (i.e. no small angle approximation) of length l in a gravitational field of strength g is given by

$$T = 4\sqrt{\frac{l}{2g}} \int_0^{\theta_0} (\cos(\theta) - \cos(\theta_0))^{-0.5} d\theta$$

where θ is the angle with the vertical, and θ_0 is the initial angle of release. Using a Monte Carlo method of integration, determine for what angles the small angle approximation is valid. (Hint: You will need to recall/research the time period for the small angle approximation and decide what is acceptable as valid.)

3. Confirm the random walk plots presented in the 'Simulation' section. Can the noise in these results be reduced? Are there any analytical results that the mean free path of a particle, undergoing random scattering, is equal to the square root of the number of scattering events?

4. Using the *Monte_Decay.f90* program:
    a. For large N (> 1000) check that N is proportional to the actual decay rate, i.e. the number of decay events per time step.
    b. Produce a plot (hint: log) that shows that spontaneous decay looks initially exponential-like but as N decreases looks increasingly stochastic in behaviour. Approximately determine where the behaviour changes. As a check, is the slope equal to λ?
    c. Is this change in behaviour independent of the initial number of atoms, and the decay constant used?

9 Partial Differential Equations

"Those are my principles, and if you don't like them... well, I have others."

— Groucho Marx

Most of the interesting equations in physics are Partial Differential Equations (PDE). Nearly all measureable quantities in the universe us vary both in space and time; a fact that is reflected by the abundance of second order PDEs in physics that have both space and time as independent variables. General speaking we call functions whose values vary in both space and time (or with any other independent variable) a field. Name any topic in physics and it will likely have a PDE describing it's phenomena; examples include but are not limited to Poisson's equation, the diffusion equation, the wave equation, the Helmholtz equation, the continuity equation, the Navier-Stokes equation, and the Schrödinger wave equation.

To solve PDEs analytically we have to employ specific techniques such as the separation of variables or through the application of the Fourier Series. Where this is difficult or not even possible, the problem might be simplified, or special cases considered, whereby the equations can be reduced to an ordinary form. However, the most general approach to solving PDEs is through the use of numerical methods.

In order to develop these numerical methods for solving PDEs we must first return to ODEs and show how to write these as *finite difference equations*. But before that, and so that everybody is on the same page, let's first discuss some generalities of PDEs ...

9.1 Classes, Boundary Values, and Initial Conditions

You may have read, or been told, that PDEs come in three different classes namely elliptic, parabolic, and hyperbolic. For second order PDEs these names are analogous to the conic sections of the same name, and are in reference to the properties of their solutions. Figure 9.1 illustrates the conic sections.

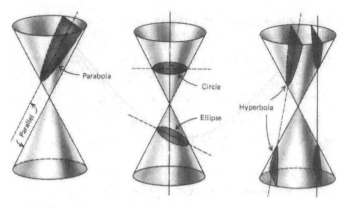

Figure 9.1: Conic sections: Parabolic, circular, elliptical, and hyperbolic. [Image copied from: http://www.andrews.edu/~calkins/math/webtexts/numb19.htm]

Just as an *ellipse* is a smooth, rounded shape, solutions to *elliptic* PDEs also tend to be smooth and rounded. Elliptic PDEs generally arise from physical problems that involve diffusion processes that have reached some equilibrium, for example a steady state temperature distribution. The *hyperbola* is the disconnected conic section. By analogy, *hyperbolic* PDEs are able to deal with discontinuities in the solution, for example a shock wave or pulse, or some instantaneous increase in temperature. Hyperbolic PDEs usually occur in relation to mechanical oscillations, such as a vibrating string. Mathematically, *parabolic* PDEs serve as a shift from the hyperbolic PDEs to the elliptic PDEs. Physically, parabolic PDEs crop up in time dependent diffusion problems, such as the transient flow of heat in a conductor, say, before it reaches a steady state.

How then do we identify a (second order) PDE as being elliptic, parabolic, or hyperbolic? Consider the most general form for a second order PDE with two independent variables

$$A\frac{\partial^2 U}{\partial x^2} + 2B\frac{\partial^2 U}{\partial x \partial y} + C\frac{\partial^2 U}{\partial y^2} + D\frac{\partial U}{\partial x} + E\frac{\partial U}{\partial y} + FU = G, \quad (9.1)$$

where A through G are arbitrary functions* of the variables x and y, and U is some physical field. Note that the second term has a factor of 2 as for any partial derivative

$$\frac{\partial^2 U}{\partial x \partial y} = \frac{\partial^2 U}{\partial y \partial x}, \quad (9.2)$$

where the left hand term is the second ordered derivative taken with respect to x first then y, and the right hand term is the second ordered derivative taken with respect to y first then x. Remember that with partial differentiation all other independent variables are considered constant when performing the operation with respect to a particular variable.

We can define the discriminant of Equation (9.1) as the following

$$d \equiv B^2 - AC. \quad (9.3)$$

When $d < 0$ the PDE is elliptic, $d = 0$ the PDE is parabolic, and $d > 0$ the PDE is hyperbolic. To see how this works let's look at some well known PDEs of different classes.

Poisson's equation in two dimensions is given by

$$\nabla^2 \varphi(x, y) = -4\pi\rho(x, y), \quad (9.4)$$

where $\nabla^2 \equiv \Delta$ is the Laplace operator that has the form

$$\Delta = \sum_i \frac{\partial^2}{\partial x_i^2}, \quad (9.5)$$

---

* In a less general case consider them constants.

with x_i representing the spatial coordinates. Here φ is a (scalar) electrical potential field and ρ is a charge density. Comparing Equation (9.4) to (9.1) we see that $A = C = 1$ and $B = 0$, thus the discriminant is negative and Poisson's equation is elliptic.

The heat equation (or more generally the diffusion equation), in one spatial dimension, is given by

$$\alpha \frac{\partial^2 T(x,t)}{\partial x^2} = \frac{\partial T(x,t)}{\partial t}, \qquad (9.6)$$

where T is some temperature field and α is known as the diffusion constant. In the case of heat flow $\alpha = K/C\rho$, where K is the thermal conductivity, C is the specific heat, and ρ is the density of the material through which the heat flows. Here $A = \alpha$ and $B = C = 0$ thus the discriminant is zero and the heat equation is parabolic.

The wave equation, in one spatial dimension is given by

$$c^2 \frac{\partial^2 \psi(x,t)}{\partial x^2} = \frac{\partial^2 \psi(x,t)}{\partial t^2}, \qquad (9.7)$$

where ψ represents the displacement of the wave from some equilibrium, say, and c is the speed of the wave. Here $A = c^2$, $C = -1$, and $B = 0$ making the discriminant positive and thus the wave equation is hyperbolic.

As with ODEs we need to know some initial values in order to determine a particular or unique solution to the PDE we are studying. For example, to obtain a particular solution of the (second order) ODE governing simple harmonic motion we needed to know the initial position and the initial velocity of the body. For second order PDEs we still require two pieces of information, but in this case they are the initial state or condition of the entire system, and the behaviour of the system at its boundaries. To illustrate this concept, imagine a metal rod that is held at a constant 0°C at one end, and 100°C at the other, i.e. we know the boundary values. Do we have sufficient information to determine the temperature distribution of the rod as it evolves with time? We can certainly guess the steady state distribution, it will be linear over the length of the rod increasing from 0°C to 100°C. However, in order to determine the transient behaviour we need to know the initial temperature distribution of the rod, i.e. its initial condition. Generally, we write the initial condition as

$$U(x_i, t=0) = U_0(x_i), \tag{9.8}$$

where U is some physical quantity, x_i represent spatial coordinates, t represents time, and it follows that U_0 is the initial state of the system.

The type of *boundary conditions* that we have used in the example above are known as *Dirichlet* conditions; the *value of the function* (temperature in our example) at the boundaries of the system domain (the ends of the rod) is known. Mathematically we write Dirichlet conditions as

$$U(x_i = \Omega_i, t) = f, \tag{9.9}$$

where f represents the value of the quantity U on the boundary Ω_i.

There are other types of boundary conditions. *Neumann* boundary conditions are when we know the *normal derivative of the function* at the boundaries. In the example of heat conduction through a rod Neumann boundary conditions relate to the flux or heat flow across the ends of the rod. Mathematically we write Neumann conditions as

$$\left.\frac{\partial U(x_i, t)}{\partial x_i}\right|_{x_i = \Omega_i} = g, \tag{9.10}$$

where g represents the flux or flow of quantity U across some boundary $x_i = \Omega_i$. Note that in the most general cases f and g can be known functions of x_i and t but to keep things relatively simple we will only consider the case where they are constants. For example, if heat flows across one end of our metal rod then physics tells us there must be a temperature gradient across that boundary i.e. g is some nominal, constant value with units of °C/m. Additionally, the sign of g tells us whether the heat flows into or out of the rod.

Cauchy boundary conditions are when we know both the value of the function and its normal derivative on the *same* boundary. For some problems a Cauchy boundary condition will *over specify* the PDE and no unique solution will exist. As a rule of thumb Cauchy conditions should only be used with hyperbolic PDEs (like the wave equation), whereas Dirichlet and Neumann conditions should only be used with elliptic and parabolic PDEs. Note, however, that the precise boundary conditions will depend on the physics being modelled, and PDE we have to solve. Sometimes this requires that we have *mixed* boundary conditions; *different boundary conditions* are used on *different parts* of the boundary of the system. For example, if our metal rod is held at a constant temperature at one end, Dirichlet condition, but is insulated at the other end, Neumann condition (g = 0), then we have mixed conditions.

Before we leave this discussion of the generalities of PDEs I just want to mention some convenient shorthand notation for expressing partial derivatives. We make the following adjustments

$$\frac{\partial U(x_i,t)}{\partial x_i} = U_{x_i}, \tag{9.11}$$

$$\frac{\partial U(x_i,t)}{\partial t} = U_t, \tag{9.12}$$

$$\frac{\partial^2 U(x_i,t)}{\partial x_i^2} = U_{x_i x_i}, \tag{9.13}$$

and so on. Although we rarely come across the mixed, second ordered derivative, for completeness

$$\frac{\partial^2 U(x_i,t)}{\partial x \partial t} = U_{x_i t}, \tag{9.14}$$

and remember that $U_{xt} = U_{tx}$.

Thus, Equation (9.1) at the start of this section can be rewritten as

$$A U_{xx} + 2 B U_{xy} + C U_{yy} + D U_x + E U_y + F U = G. \tag{9.15}$$

9.2 Finite Difference Methods

We saw in Chapter 5 an approach to solving second order ODEs that involved generating an auxiliary variable, and separating the ODE into a pair of, coupled first order ODEs. While this perfectly good method for many equations found in physics it can be difficult, if not impossible, to apply to PDEs.

9.2.1 Difference Formulas

Another approach to solving boundary value problems is to approximate the differential equation with a difference equation. For now we will only consider differential equations with one independent variable, i.e. ODEs, and will later show how to extend this to PDEs. We can obtain the difference equation by considering the definition of the differential equation, which is

$$\frac{\Delta f(x)}{\Delta x} = \frac{f(x+h)-f(x)}{h} \underset{\lim h \to 0}{=} \frac{df(x)}{dx}. \tag{9.16}$$

Thus an appropriate approximation to the derivative is

$$f'(x) \approx \frac{f(x+h)-f(x)}{h}, \tag{9.17}$$

where h is some small but finite value.

To determine the error behaviour of this approximation we could compare it to the analytical solution of some known ODE. However, we can do a much better job by using the mathematical tools at our disposal. As some of you may have already guessed, we can obtain Equation (9.17) using the Taylor series expansion of a function at $f(x+h)$ about $f(x)$ i.e.

$$f(x+h) = f(x) + hf'(x) + \frac{h^2}{2!}f''(x) + \cdots. \tag{9.18}$$

This can rearrange for the first order derivative such that

$$f'(x) = \frac{1}{h}[f(x+h) - f(x) - \frac{h^2}{2!}f''(x) - \cdots]. \qquad (9.19)$$

Notice the equivalency; Equation (9.19) is not an approximation but an exact formula for the first ordered derivative. Comparing Equation (9.17) with Equation (9.19) we see that the approximation for the derivative is actually the Taylor series expansion truncated after the second term. Therefore the error in the approximation is $\mathcal{O}(h)$ (note the multiplicative factor of $1/h$) and we can write the equivalency

$$f'(x) = \frac{f(x+h) - f(x)}{h} + \mathcal{O}(h). \qquad (9.20)$$

This is called the forward difference approximation to the derivative as we are using a point that is forward one step, $f(x+h)$, to approximate the derivative at the current position, $f(x)$.

A backward difference is derived similarly such that

$$f'(x) = \frac{f(x) - f(x-h)}{h} + \mathcal{O}(h), \qquad (9.21)$$

where we use a step behind, $f(x-h)$, to approximate the derivative at the current position, $f(x)$. This also has $\mathcal{O}(h)$ accuracy. For convenient notion let's make the following changes: if our current position is given by $f(x) \rightarrow f(x_i) \rightarrow f_i$, the forward position is then given by $f(x+h) \rightarrow f_{i+1}$, and the backward position is given by $f(x-h) \rightarrow f_{i-1}$. Both the forward and backward difference approximations are two-point formulas.

An advantage of deriving these expressions from their respective Taylor series is that we notice the leading error term in both has the same magnitude but opposite sign. Thus if we add Equations (9.20) and (9.21) this leading error term will cancel, and we should obtain a more accurate approximation.

After performing the necessary steps we find that

$$f'_i = \frac{f_{i+1} - f_{i-1}}{2h} + O(h^2). \tag{9.22}$$

This is called the central difference formula, and is a three-point formula as we use both the forward, $i + 1$, and backward, $i - 1$, values to approximate the derivative at our current value, i. You can think of the formula as containing an f_i term but with a zero coefficient. Note that this formula takes a symmetrical "picture" at the local neighbourhood of the point of interest, whereas the forward and backward formulas only use the information to one side of the point of interest.

We could keep improving the error behaviour by taking more points around our point of interest. For example a five-point formula can be derived in a similar manner to the three-point formula above yielding

$$f'_i = \frac{1}{12h}[f_{i-2} - 8f_{i-1} + 8f_{i+1} - f_{i+2}] + O(h^4). \tag{9.23}$$

Notice that we have an error behaviour of $O(h^4)$. In general, an n point central difference formula will be $O(h^{n-1})$ accurate. While we could just keep increasing the number of points in the approximation for the derivative to improve its accuracy this does have a practical limit. These approximations require that we know values for the function both ahead and behind our current position. Unlike, the Euler or Runge-Kutta methods that are self-starting formulas, multi-step methods require that we initiate them in some way. Typically, this involves using a Runge-Kutta method, say, to provide the required number of points from the boundary conditions to start the multi-step formula. As any error incurred during this initiation stage will be propagated throughout the integration, the initiation method needs to be at least the same order of accuracy as the multi-step formula it is starting. As the values of the function and its derivative(s) on and close to the boundaries of a real physical system are usually important (if not crucial) to the outcome of the integration, they must be calculated accurately.

Note that should we be approximating the derivatives of a *known function* on a particular interval we can use an interpolation method to *extrapolate* the derivatives at the limits (boundaries) of the interval. For example, let's imagine we have some function, $f(x)$, stored in an array of size N such that we have function values at equidistance increments, h, over x. We then approximate the first ordered derivative using the five-point formula, say. Necessarily, we have to start the approximation at array element 3 and end at array element $N - 2$ to account for the number of points required behind and in front of the point of interest. To obtain the missing values for the derivative we extrapolate from those we have calculated in the main body of the array. As we have used a five-point finite difference method we should use an interpolation method that matches the accuracy order, for example a four point (3rd ordered) Lagrange polynomial interpolation scheme which has a quartic error behaviour.

This is not the only method we could use. As we known the function values we could apply a forward and a backward difference formula at the start and at the end of our array, respectively, to approximate the derivative(s) at those points. Keep in mind that these methods to deal with edge or boundary values such match the error behaviour of the method applied to the interior values of the function.

The more practical (and obvious) way of increasing the accuracy of our approximation formulas is to decrease the step size h. Of course this requires more computational effort but as with most things there is always a cost-reward trade off. And, as we shall see shortly, knowing how the error behaves as the step size decreases can be used to our benefit. That said what happens if we let $h \to 0$ in a computer program that calculates finite differences?

Higher order differential equations can also be approximated with a difference equation. Again they are derived using the Taylor series expansion of various points about the point of interest. The central difference formula for the second ordered derivative is

$$f_i'' = \frac{f_{i+1} - 2f_i + f_{i-1}}{h^2} + O(h^2), \qquad (9.24)$$

which is a three-point formula. The five point central difference formula for the second ordered derivative is given by

$$f_i'' = \frac{1}{12h^2}[-f_{i-2} + 16f_{i-1} - 30f_i + 16f_{i+1} - f_{i+2}] + \mathcal{O}(h^4). \quad (9.25)$$

9.2.2 Application of Difference Formulas

How do we go about applying these formulas to a particular differential equation? To illustrate this consider the following general, second order ODE

$$\alpha f'' + \beta f' + \gamma f = \delta x, \quad (9.26)$$

where α through δ are constants, and is subject to the boundary conditions

$$f(a) = c, \quad f(b) = d, \quad (9.27)$$

where $[a, b]$ defines the computational domain (integration interval), and c and d are the values of the function at the boundaries (which are what kind of boundary conditions?).

The first step is to partition our computational domain into a grid or mesh of discrete points. For simplicity we make this grid uniform by defining

$$h = \frac{b-a}{N-1}, \quad (9.28)$$

where N is the *total* number of grid points; we include both the boundaries in our grid*. We then approximate the continuous, differential equation with a discrete, difference equation to find the solution on the grid we have just imposed. Substituting the three point central difference formulas into Equation (9.26) we arrive at the following finite difference equation for some arbitrary interior point x_i

$$\frac{\alpha}{h^2}(f_{i+1} - 2f_i + f_{i-1}) + \frac{\beta}{2h}(f_{i+1} - f_{i-1}) + \gamma f_i = \delta x_i. \quad (9.29)$$

---

* e.g. $a = 0$, $b = 1$, with $h = 0.5$, hence $N = 3$; grid points at $x = 0, 0.5,$ and 1.

We then solve Equation (9.29) on the $N-2$ interior grid points; the boundary values are fixed, i.e. $f(a) \equiv f_1 = c$, and $f(b) \equiv f_N = d$. This means we have a system of $N-2$ linear equations that can be solved simultaneously for the $N-2$ unknown interior grid points. Note that the solution on the discrete grid will only approximate the solution of the original, continuous problem. That, of course, is the point; if we had access to an exact, analytical solution to the problem we shouldn't need to approximate the problem in the first place, but I digress.

There are now two ways to proceed with the solution of Equation (9.29). We either go for the direct method or the indirect method. The direct method involves recasting Equation (9.29) in matrix form and solving the system by Gaussian elimination, say. To do this we gather like terms such that

$$\varphi f_{i+1} + \theta f_i + \psi f_{i-1} = \delta x_i, \qquad (9.30)$$

where

$$\varphi = \left(\frac{\alpha}{h^2} + \frac{\beta}{2h} \right),$$

$$\theta = \left(\gamma - \frac{2\alpha}{h^2} \right),$$

and

$$\psi = \left(\frac{\alpha}{h^2} - \frac{\beta}{2h} \right).$$

Perhaps not immediately obvious but Equation (9.30) has the matrix form of

$$A\underline{f} = \delta\underline{x}, \qquad (9.31)$$

where A is an $(N-2)$-by-$(N-2)$ tridiagonal matrix given by

$$A = \begin{bmatrix} \theta & \varphi & 0 & \cdots & 0 \\ \psi & \theta & \ddots & \ddots & \vdots \\ 0 & \ddots & \ddots & \ddots & 0 \\ \vdots & \ddots & \ddots & \theta & \varphi \\ 0 & \cdots & 0 & \psi & \theta \end{bmatrix}; \qquad (9.32)$$

$\underline{f}$ is an $N - 2$ vector representing the unknown function values at the interior grid points, explicitly

$$\underline{f} = \begin{bmatrix} f_2 \\ f_3 \\ \vdots \\ f_{N-2} \\ f_{N-1} \end{bmatrix}; \qquad (9.33)$$

and $\underline{x}$ is an $N - 2$ vector representing the known, discrete grid of points on the independent variable x, explicitly

$$\underline{x} = \begin{bmatrix} x_2 \\ x_3 \\ \vdots \\ x_{N-2} \\ x_{N-1} \end{bmatrix}, \qquad (9.34)$$

where $x_i = a + (i - 1)h$ for $i = 2, \ldots, N - 1$.

Equation (9.31) can be readily solved by the LAPACK subroutine 'DGTSV', say, which solves a general tridiagonal system of linear equations by Gaussian elimination with partial row pivoting.

The indirect method of solution is to find an equation for f_i in terms of its neighbouring points and find an approximation that satisfies the difference equation at the grid points using an iterative technique. That is, we start with an initial guess for f at all grid points and using the appropriate equation we iterate to a more accurate approximation (we hope).

Partial Differential Equations

We express our iterative scheme by rewriting Equation (9.30) to solve for f_i such that

$$f_i^{(n)} = -\frac{1}{\theta}\left(\varphi f_{i+1}^{(n-1)} + \psi f_{i-1}^{(n-1)} - \delta x_i\right), \qquad (9.35)$$

where n is an index that represents the level of iteration; it should not be confused with a power. Thus our initial guess is given by the index $n = 0$ and we iterate to the next level, $n = 1$, through application of Equation (9.35) at all interior grid points; the boundaries are fixed. Note that we need to have two storage arrays; one to store the grid point values at the current level of iteration, and one to store the grid point values at the next level of iteration. Note also that the scheme can be described as being red-black in reference to a chessboard pattern of the same colours. To explain this, take an even value for the grid point index i. Note that the next level of iteration depends only on the adjacent, odd values of the current iteration level; the grid points x_i remain constant throughout the iteration scheme. This means that, for each iteration level, the computations for even and odd grid points can be done independently. Thus if we think of the iteration scheme as a red and black chessboard, where the squares represent the grid points and the columns represent the iteration level, then red squares only influence other red squares, and black squares only influence other black squares. This property lends itself well to parallel computing, but more on this in a later chapter. The iteration method we have just outlined is called the *Jacobi* scheme and will converge to the exact solution.

Parallel computing aside, we can apply a little thought to the Jacobi scheme and come up with a similar method but with quicker convergence. Imagine you've just determined $f_i^{(n)}$ and are ready to compute the next grid point $f_{i+1}^{(n)}$. The Jacobi scheme would have you use the value $f_i^{(n-1)}$ at the previous iteration level despite the fact you have just calculated an improved value for that grid point. Instead, let's use that improved value. Equation (9.35) then becomes

$$f_i^{(n)} = -\frac{1}{\theta}\left(\varphi f_{i+1}^{(n-1)} + \psi f_{i-1}^{(n)} - \delta x_i\right). \qquad (9.36)$$

Note that this formula represents moving through the array with ascending i values. For descending i values then the iteration index levels on the right hand side swap accordingly. This iteration method is called the *Gauss-Seidel* scheme, and will converge to the exact solution more quickly than the Jacobi scheme. However, note that we have now lost the red-black property of the Jacobi scheme, making it more of a challenge to write the Gauss-Seidel scheme in parallel code.

Iterative methods are generally inferior to direct methods when applied to ODEs. By this we mean that direct methods can supply us with a solution with far less computational effort than that of an iterative scheme to the same level of accuracy. Iterative methods come into their own when applied to physical problems that involve several independent variables, in other words, PDEs. This is in part due to the differences in the propagation of error between the two methods. Direct methods rely on matrix factorisation, typically Gaussian elimination, with the solution found by back substitution. Any error in one value is passed to all values that follow it in the substitution, leading to a non-uniform distribution of error in the numerical solution. As we add more independent variables this accumulation of error tends to worsen. For iterative methods the error tends to get smeared out across the entire computational domain (across each independent variable) leading to a more uniform distribution. Additionally, the error in an iterative method can always be improved by simply iterating further, whereas the error in a direct method cannot be improved unless applying Richardson extrapolation (see the next section), for example.

When writing code for an iteration scheme we must bear in mind that *all* grid points must reach a specified level of accuracy before we can say the iteration has converged. That is, we check that the difference between the same grid points at different iteration levels is less than some tolerance, for all the grid points in the computational domain. This is easily accomplished by using a logical variable which is set to true at the start of each iteration, then set to false should *any* of the grid points fail the accuracy test i.e. it only takes one to fail for the convergence check to fail. If the convergence check has failed then we go to the next level of iteration. The tolerance itself should be set relatively large as the solution we obtain is only relevant to the discrete, finite difference approximation we made of the continuous, differential equation. Of course the accuracy of our approximate problem can be improved by making the grid finer, in which case we could also apply Richardson extrapolation.

Though convergence is (pretty much) guaranteed it can be slow, thus count the number of iterations and have the program exit should we go beyond some maximum; this is true of any iteration method.

The file *gauss-seidel.f90* contains the program code to implement the iteration method of the same name. You may find it useful to add some code to this program to have it print out some measure of the error between iterations should the method fail to converge before reaching the maximum iteration count. This may give you some feel for how close the method was to convergence and allow you to adjust either the tolerance or maximum iteration count appropriately.

Both the Jacobi and Gauss-Seidel schemes are what are known as relaxation techniques. That is, we derive the finite difference approximation to the differential equation, guess at a solution, and the method relaxes that guess to the exact value. How quickly the method relaxes the solution to its exact value depends upon how good the initial guess was in the first place. However, no matter how good or bad the initial guess the method will eventually converge on a significantly accurate approximation and the relaxation becomes what is called monotonic. When a function is monotonic it only ever increases or decreases with its independent variable; there are no oscillations, inflections, or stationary points. In this case the relaxation takes the solution closer to the true value at all grid points after each iteration loop. This suggests that if we take a weighted average between the iterate value we have just calculated and the previous iterate value we should obtain a more accurate solution for that grid point. Mathematically, this is

$$f_i^{(n)} = \omega \overline{f}_i^{(n)} + (1-\omega) f_i^{(n-1)}, \tag{9.37}$$

where ω is the weighting factor (sometimes referred to as the extrapolation or relaxation factor) and lies on the interval [0,2], and

$\bar{f}_i^{(n)}$ is the Gauss-Seidel value. Equation (9.37) is called *over-relaxation* and as we apply it successively at each iteration level the entire method is referred to as *Successive Over-Relaxation* (SOR).

For each individual differential equation, and hence derived finite difference approximation, there will be an optimal value for ω that gives the most rapid convergence. This optimal value can be calculated in some cases and estimated in others.

Currently we have only developed finite difference approximations to ODEs but the extension to PDEs is not so difficult. However, before we embark upon that discussion let's take a bit of a tangent and talk about ...

9.3 Richardson Extrapolation

I have made reference to Richardson extrapolation before in the sections on developing adaptive step numerical integration (quadrature) methods, the adaptive step ODE solvers, and most recently in the section above. So let's discuss the general technique and how it relates to computations of all sorts.

Richardson extrapolation is what is known as a sequence acceleration method. It is named after Lewis Fry Richardson, who introduced the technique in the 1920s for use in predicting the weather via numerical techniques. In essence the technique uses a single formula with known error behaviour, which is computed using different step sizes, and the results are combined to eliminate the leading error term in the sequence. To explain how it works in detail lets first apply it to the particular method of numerical differentiation and then generalise for any method where the error behaviour is known.

We know from the Taylor series expansion that the first ordered derivative of any function can be given by

$$f_i' = \frac{f_{i+1} - f_i}{h} - \frac{h}{2!} f_i'' - \cdots, \tag{9.38}$$

so that the approximation to the derivative is given by

$$A_0(h) = \frac{f_{i+1} - f_i}{h},\qquad(9.39)$$

with an $O(h)$ error behaviour. The index refers to the extrapolation level; this will be explained shortly. Remember that here we are using the specific example of the central difference formula. We could use any numerical technique that we have discussed thus far so long as we know its error behaviour.

If we halve the step size h then the new approximation will have a leading term error that is half that of the previous approximation. Thus we can eliminate the leading error term by subtracting the approximation using h from twice the approximation using $h/2$. Thus the extrapolated approximation is given by

$$A_1(h/2) = 2A_0(h/2) - A_0(h),\qquad(9.40)$$

which has an $O(h^2)$ error behaviour as we eliminated the leading $O(h)$ error term. If we halve the step size again and calculate $A_0(h/4)$ then we can apply the same technique to find

$$A_1(h/4) = 2A_0(h/4) - A_0(h/2),\qquad(9.41)$$

with the same error behaviour as before. Now here's the clever bit. As we know how the error behaves in the extrapolated approximations, $O(h^2)$, and we have two measures of it with differing step size we can eliminate the next leading error term by performing

$$A_2(h/4) = \frac{4A_1(h/4) - A_1(h/2)}{3}.\qquad(9.42)$$

Here we have used the fact that the leading error term in $A_1(h/4)$ must be four times smaller than the leading error term in $A_1(h/2)$. Therefore, four times $A_1(h/4)$ minus $A_1(h/2)$ must leave three times the exact answer plus the remaining error terms. The error behaviour in Equation (9.42) is then $O(h^3)$.

This process can continue indefinitely with a general recurrence relation

$$A_{m+1,l+1}(h) = \frac{t^{k_m} A_{m+1,l} - A_{m,l}}{t^{k_m} - 1}, m \leq l, \qquad (9.43)$$

where

$$A_{m,l} = A_l\left(\frac{h}{t^m}\right), \qquad (9.44)$$

l is the extrapolation level, t is the factor we use to reduce the step size h (for practical purposes this is nearly always two), and k_m is an integer related to the step size reduction level m and the nature of the sequence we are extrapolating. For instance, the central difference formula, Equation (9.22), has only error terms involving even power terms of h, thus $k_m = 2m$. Conversely, if we had a method that involved only odd power terms of h then $k_m = 2m + 1$. In the most general cases k_m need not be an integer but we won't consider those (because we're not sadists – at least I'm not).

As Equation (9.43) is somewhat abstract lets apply it to an actual example. Consider the function $f(x) = \sin(x)$ with its first ordered derivative approximated by the three-point central difference formula, Equation (9.22). We can put the approximations into matrix format with the step size reduction level, m, defining the rows and the extrapolation level, l, defining the columns; note that we start the numbering at zero. Thus we gain the following approximation matrix for the first two levels of extrapolation:

$$A = \begin{bmatrix} 0.5260090707 & 0.0000000000 & 0.0000000000 \\ 0.5367074877 & 0.5402736266 & 0.0000000000 \\ 0.5394022522 & 0.5403005070 & 0.5403022990 \end{bmatrix}. \qquad (9.45)$$

Here we assessed the derivative at $x = 1$, with an initial $h = 0.4$, $t = 2$, and we use the fact that $k_m = 2m$. The way this matrix equation fills is row by row. In other words we compute a new level of extrapolation as soon as we have sufficient information to do so. For example, we first compute $A_{0,0}$ and $A_{1,0}$, then use those to calculate $A_{1,1}$, before returning to the zeroth level of extrapolation to calculate $A_{2,0}$, and the procedure continues. In this way the bottom right most entry of A is the most accurate approximation to the derivative. The matrix A is called a lower triangular matrix for obvious reasons.

The first question you should be asking yourselves is "are we actually saving ourselves computational effort?". If say instead of approximating a derivative via a finite difference we were performing a numerical quadrature, such as the composite trapezoidal rule, then yes we do save ourselves computational effort. To explain, the first column of A would contain the approximations from the trapezoidal rule calculated at the different step sizes (slice width). If the first approximation took N function evaluations then to calculate the first column of A up to the second level of step size reduction would require $N + 2N + 4N = 7N$ function evaluations. Using these approximations we can obtain a more accurate value up to the second level of extrapolation. At this level the error behaves as $\mathcal{O}(h^6)$; the trapezoidal rule has no odd powers of h in its error terms. Thus the error in the approximation at this level will be roughly a factor of $1/4^6$ times smaller than the first entry. To obtain this accuracy order by only reducing the step size would require us to use at least $1/64$ times the original step size! In other words $64N$ function evaluations versus 7N function evaluations. The further we extrapolate the bigger (and the better) the difference between these two numbers.

However, for any finite difference formula we only ever need to use a set number of evaluations to compute an approximation to the derivative at a given point. For instance the three-point central difference formula requires two function evaluations*, one at f_{i+1} and one at f_{i-1}, regardless of the step size. The extrapolation would appear to be wasting computational effort. However, if we have an unknown function taken from some measurement, say, and we wish to find its derivative we may not have enough points to analyse the

---

* Remember that here the f_i term has a coefficient equal to zero.

absolute error in our approximation using different step sizes of the central difference formula alone. Richardson extrapolation would (hopefully) converge more rapidly on a precise approximation, and give us means to estimate the absolute error in the numerical derivative. Secondary to this point is that as we reduce the step size the function evaluations for the finite difference come closer together in size. Eventually, their difference becomes comparable to the machine precision and we introduce round-off error. You can see this for yourselves by applying the central difference formula to a known function and calculating the absolute error in the approximation for a decreasing step size. You should find that the error reduction initially behaves as predicted but will eventually slow down and even begin to increase for sufficiently small step sizes due to the round-off error. By using Richardson extrapolation we shouldn't run into these round-off error issues.

Typically, the extrapolation doesn't go much beyond the third or fourth level as by then we usually have an approximation that is within a desired tolerance level. One way to insure the significant figure of accuracy of the approximation is to subtract the previous level of extrapolation from the current level (at the same step size reduction level, i.e. the same row) and test it against a desired tolerance. For instance, in our example above subtracting element $A_{2,1}$ from $A_{2,2}$ we would seen that $A_{2,2}$ is at least 5 significant figures accurate i.e. within the tolerance that we used of 10^{-4}. In fact the approximation shown is 6 significant figures accurate.

The file *richardsonExtrap.f90* contains the code to perform the extrapolation using a matrix array A and a three-point central difference formula to approximate the first ordered derivative of some user defined function. You could of course swap the central difference subroutine for any numerical method of your choice. Though beware that the extrapolation has been written specifically for numerical methods that contain no odd powers of h in their error terms. How might you modify this so that it could cope with any general numerical method?

With that aside over lets return to the topic at hand ...

9.4 Numerical Methods to Solve PDEs

To begin this discussion let's start with probably the most intuitive of the partial differential equations, namely....

9.4.1 The Heat Equation with Dirichlet Boundaries

Consider the normalized (i.e. the diffusion constant is unity) heat equation in one dimension across a metal rod

$$U_t = U_{xx}, \tag{9.46}$$

where U is the temperature field, x is the spatial coordinate, and t is time. Lets impose the Dirichlet boundary conditions

$$U(0,t) = U(L,t) = c, \tag{9.47}$$

where L is the length of the rod, and c is a constant temperature. This is the model for a metal rod held at a steady temperature at either end. The initial condition of the rod is given by

$$U(x,0) = U_0(x). \tag{9.48}$$

One way to solve this equation numerically is to approximate all the derivatives by finite differences. We partition the domain (i.e. the length of the rod) in space using a grid or mesh $x_1, \ldots, x_M$ (hence M spatial grid points in total) and in time using a grid $t_1, \ldots, t_N$ (hence N time grid points in total). We assume a uniform partition both in space and in time such that the difference between two consecutive space points can be given by h, and between two consecutive time points can be given by k. The points

$$u(x_m, t_n) = u_m^n \tag{9.49}$$

represent the numerical approximation to the exact solution at the grid point (m, n). To avoid confusion a lower case letter with two indices either written u_m^n or $u_{m,n}$ refer to the *grid points* of the *finite difference approximation*, the former being for one spatial variable and a time variable, and the latter being used for two spatial variables. An upper case letter followed by two indices without a comma, e.g. U_{xx}, is the second ordered partial derivative of U with respect to the continuous variable x. Here I'm using the convention that we write the discrete time point index as a superscript; we are not taking a power. For clarity, if we had a two dimensional heat conducting sheet, say, and we partitioned the sheet across x, y, and t then the conventional notation for a grid point on the temperature field U would be

$$u(x_l, y_m, t_n) = u_{l,m}^n, \qquad (9.50)$$

where l and m are the indices for the spatial coordinate grid points, and n is the index for the temporal grid points.

Explicit method

Using a forward difference at time t_n and a second-order central difference for the space derivative at position x_m we get the recurrence equation:

$$\frac{u_m^{n+1} - u_m^n}{k} = \frac{u_{m+1}^n - 2u_m^n + u_{m-1}^n}{h^2}. \qquad (9.51)$$

This is an explicit method for solving the one-dimensional heat equation in that we can compute the advanced time step u_m^{n+1} from the previous values such that

$$u_m^{n+1} = (1 - 2r)u_m^n + ru_{m-1}^n + ru_{m+1}^n, \qquad (9.52)$$

where $r = k/h^2$.

To input the boundary conditions we have to fix $u_0^n = c$ and $u_M^n = c$, over all N time steps.

To visualise Equation (9.52) we can consider what is called the computational or numerical stencil[*] of the explicit method. Figure 9.2 shows this stencil placed at some arbitrary interior location, i.e. not near the boundaries of the domain. Here we use the notion that known values, either computed or belonging to the initial or boundary conditions, are represented by filled squares, and the

---

[*] Sometimes, and rather grandiosely, this is called a molecule rather than a stencil.

unknown values, those yet to be computed, are represented by open circles. The stencil shows us that we are using three known values, specifically u_{m-1}^n, u_m^n, and, u_{m+1}^n to compute the unknown value u_m^{n+1}. As we run the computation we can think of the stencil as moving down one row as a time, calculating the unknown value at the advance time step, until it reaches the edge of the grid (the domain boundary), where it moves to the top of the next column and repeats the process, finishing at the bottom of the last time step.

The explicit method is known to be numerically stable and convergent when $r \leq 1/2$. As we have taken a forward difference for the time differential, and a central difference for the space differential the error in the approximation can be written as $\Delta u = \mathcal{O}(k) + \mathcal{O}(h^2)$.

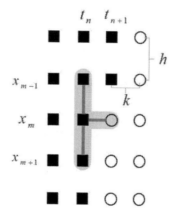

Figure 9.2: Explicit method stencil. Filled squares are computed points; open circles are to be computed.

Implicit method

If we use the backward difference at time t_{n+1} and a second-order central difference for the space derivative at position x_m we get the recurrence equation

$$\frac{u_m^{n+1} - u_m^n}{k} = \frac{u_{m+1}^{n+1} - 2u_m^{n+1} + u_{m-1}^{n+1}}{h^2}. \tag{9.53}$$

This is an implicit method for solving the one-dimensional heat equation in that to obtain the advanced time step t_{n+1} we compute u_m^{n+1} by solving a system of linear equations. We obtain that system of linear equations by rearranging Equation (9.53) to give

$$(1+2r)u_m^{n+1} - ru_{m-1}^{n+1} - ru_{m+1}^{n+1} = u_m^n. \tag{9.54}$$

The numerical stencil for the implicit method is shown in Figure 9.3. As we can see each equation contains one known value u_m^n and three unknown values u_{m-1}^{n+1}, u_m^{n+1}, and u_{m+1}^{n+1}. If there's a thought scratching at the back of your head then remember that we move this stencil down one row at a time. Thus, for the grid points $m = 3, \ldots, M - 2$ each unknown appears in three separate equations, hence we have sufficient information to compute the unknowns at the advanced time step for those grid points. For instance, in Figure 9.3 we now have sufficient information to calculate grid point u_{m-1}^{n+1}. But what of grid points $m = 2$, and $m = M - 1$? They only appear in two separate equations; it looks as if we are missing two known values, one for each grid point! Of course the answer lays in the boundary values. If we place our implicit method stencil at $m = 2$, say, then one of the unknowns is actually a boundary value, which is a known value. The same can be said at $m = M - 1$. Thus we have sufficient information to calculate u_2^{n+1} and u_{M-1}^{n+1}. We'll discuss how to deal with these boundary values in detail shortly.

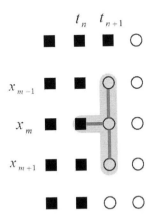

Figure 9.3: Implicit method stencil

The implicit method is always numerically stable and convergent but is typically more computationally intensive than the explicit method as it requires solving a system of numerical equations at each time step. As we have used the backward time difference and the central space difference the (local) error in the approximation is the same as the explicit method. Note that the global error may be different due to the different methods of computation required to get to the advanced time step.

Crank–Nicolson method

Finally if we use a central difference approximation at time $t_{n+1/2}$ and a second-order central difference for the space derivative at position x_m we get the recurrence equation:

$$\frac{u_m^{n+1} - u_m^n}{k} = \frac{1}{2}\left(\frac{u_{m+1}^{n+1} - 2u_m^{n+1} + u_{m-1}^{n+1}}{h^2} + \frac{u_{m+1}^n - 2u_m^n + u_{m-1}^n}{h^2} \right). \quad (9.55)$$

Remember to get the central difference formula we sum the forward difference with the backward difference; the factor of one half takes into account that we are assessing the time half way between grid points.

Equation (9.55) is known as the Crank–Nicolson method. The method was developed by two Britons namely John Crank, a mathematical physicist, and Phyllis Nicolson a mathematician, in the mid-20th century. Note that this method is implicit as the approximation to the advanced time step is dependent on itself and we can obtain u_m^{n+1} from solving the system of linear equations given by

$$2(1+r)u_m^{n+1} - r\left(u_{m-1}^{n+1} + u_{m+1}^{n+1}\right) = 2(1-r)u_m^n + r\left(u_{m-1}^n + u_{m+1}^n\right). \quad (9.56)$$

The stencil for the Crank-Nicolson method is shown in Figure 9.4. Here we use three known values and three unknown values to generate an equation.

Note that this stencil is still only moved down one row at a time. In its current position in Figure 9.4 we have enough information to compute grid point u_{m-1}^{n+1}.

The Crank-Nicolson method is always numerically stable and convergent. However be aware that if $r > 1/2$ this method may produce unwanted decaying oscillations in the solution. The errors for the Crank-Nicolson method are quadratic over both the time step k, and the space step h.

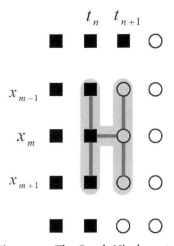

Figure 9.4: The Crank-Nicolson stencil.

General Finite Difference Method

We can condense these three methods into one elegant formula through the use of a so called weighted variable θ. Using this variable we can rewrite the finite difference approximations above as

$$\frac{u_m^{n+1} - u_m^n}{k} = \theta \frac{u_{m+1}^{n+1} - 2u_m^{n+1} + u_{m-1}^{n+1}}{h^2} + (1-\theta) \frac{u_{m+1}^n - 2u_m^n + u_{m-1}^n}{h^2}, \quad (9.57)$$

where $\theta = 0$ gives the explicit method, $\theta = 1/2$ gives the Crank-Nicolson method, and $\theta = 1$ is the implicit method, though θ could have any value in the range [0,1].

Rearranging Equation (9.57) into an expression for the advanced time step we obtain

$$-(r\theta)u_{m+1}^{n+1} + (1+2r\theta)u_m^{n+1} - (r\theta)u_{m-1}^{n+1}$$
$$= r(1-\theta)u_{m+1}^n + (1-2r(1-\theta))u_m^n + r(1-\theta)u_{m-1}^n. \qquad (9.58)$$

Note that Equation (9.45) is for the *interior* nodes of the domain grid, i.e. $m = 2, \ldots, M - 1$. The values of the grid nodes u_1^n and u_M^n are given by the boundary conditions.

Although not obvious at all Equation (9.58) represents a tridiagonal system of linear equations, i.e. it can be written in the form of

$$A\underline{u}_{n+1} = B\underline{\tilde{u}}_n + \Omega_{n+1}, \qquad (9.59)$$

where A is an $(M-2)$-by-$(M-2)$ tridiagonal matrix given by

$$A = \begin{bmatrix} 1+2r\theta & -r\theta & 0 & \cdots & 0 \\ -r\theta & 1+2r\theta & -r\theta & \ddots & \vdots \\ 0 & -r\theta & \ddots & \ddots & 0 \\ \vdots & \ddots & \ddots & \ddots & -r\theta \\ 0 & \cdots & 0 & -r\theta & 1+2r\theta \end{bmatrix}; \qquad (9.60)$$

$\underline{u}_{n+1}$ is an $M-2$ sized vector of the function values at the interior grid nodes at the advanced time step given by

$$\underline{u}_{n+1} = \begin{bmatrix} u_2 \\ u_3 \\ \vdots \\ u_{M-2} \\ u_{M-1} \end{bmatrix}_{n+1}; \qquad (9.61)$$

$\underline{\tilde{u}}_n$ is an M sized vector of the function values at the current time step, i.e. the known values, given by

$$\tilde{u}_n = \begin{bmatrix} u_1 \\ u_2 \\ \vdots \\ u_{M-1} \\ u_M \end{bmatrix}_n ; \qquad (9.62)$$

B is an $(M - 2)$-by-M tridiagonal-ish (technical term) matrix given by

$$B = \begin{bmatrix} r\varphi & 1-2r\varphi & r\varphi & 0 & \cdots & 0 & 0 \\ 0 & r\varphi & 1-2r\varphi & r\varphi & \ddots & \vdots & \vdots \\ \vdots & 0 & r\varphi & \ddots & \ddots & 0 & \vdots \\ \vdots & \vdots & & \ddots & \ddots & r\varphi & 0 \\ 0 & 0 & \cdots & 0 & r\varphi & 1-2r\varphi & r\varphi \end{bmatrix}, \qquad (9.63)$$

where for convenience $\varphi = (1 - \theta)$; and Ω is a vector of size $M - 2$ to deal with the boundary conditions given by

$$\Omega_{n+1} = r\theta \begin{bmatrix} u_1 \\ 0 \\ \vdots \\ 0 \\ u_M \end{bmatrix}_{n+1}, \qquad (9.64)$$

i.e. there are $M - 4$ zeros between the first and last elements. Note that we have shifted our time index to being a subscript for the matrix-vector equations – just to keep you on your toes.

To explain Ω, consider Equation (9.45) at grid point $m = 2$, we write

$$\begin{aligned} -(r\theta)u_3^{n+1} + (1+2r\theta)u_2^{n+1} - (r\theta)u_1^{n+1} \\ = r(1-\theta)u_3^n + (1-2r(1-\theta))u_2^n + r(1-\theta)u_1^n \end{aligned}. \qquad (9.65)$$

Remember that we should write these equations with the unknown values on the left hand side and the known values on the right hand side. The value u_1^{n+1} is the boundary value at the advanced time step.

As the boundary is a known value it should be moved to the right hand side of the equation. A similar argument can be made for the grid point $m = M - 1$ with the boundary value u_M^{n+1} also being moved to the right hand side. Note that this manipulation is particular to Dirichlet boundary conditions. For other types of boundary conditions different methods of dealing with these boundary values have to be applied. We will discuss how to deal with a Neumann boundary condition in the next section.

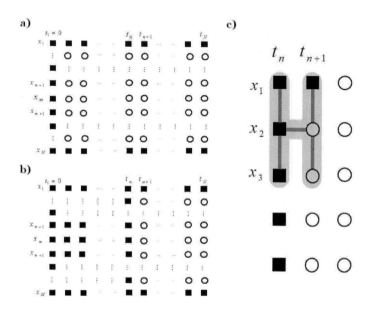

Figure 9.5: (a) The initial state of the system; (b) The state of the system at time t_n; (c) the general stencil applied at a boundary.

If the source of these manipulations is unclear from the equations then consider Figure 9.5(a). This shows the space-time domain which we have discretised into a grid of points; the space axis is vertical and the time axis is horizontal. Thus, each column of points represents the solution across the spatial dimension at each time step we are taking. The open circles represent the unknown, interior grid points that we are attempting to calculate, i.e. the left hand terms of our equations, whereas the filled squares represent our known values, i.e. the right hand terms of our equations. The first and last *rows* of the grid are given by the *boundary conditions*, whereas the first *column* is given by the *initial condition* of the function. Figure 9.5(b) shows the state of the grid at the nth time step. Here n columns have changed from unknown values to known values and we are ready to calculate the $(n+1)^{th}$ time step. If we apply our Crank-Nicolson stencil to the grid point u_2^{n+1}, Figure 9.5(c), then we see that it contains a boundary value at the advanced time step. Necessarily, this value has to be brought to the right hand side of our equations; hence the value appearing at the start of Ω. Similarly, the stencil applied to grid point u_{M-1}^{n+1} also includes a boundary value, hence the value at the end of Ω.

It was mentioned earlier that we would only deal with Dirichlet conditions that are constant values, and not functions of time (or other spatial coordinates). The reason for this becomes apparent when we consider the addition of Ω in Equation (9.59). As $u_1^n = u_1^{n+1}$ and $u_M^n = u_M^{n+1}$ for all n, we can incorporate Ω into matrix B such that the first and last entries of B both simplify to r.

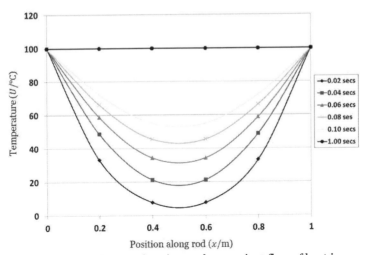

Figure 9.6: Numerical approximation to the transient flow of heat in a one dimensional metal rod with Dirichlet boundary conditions.

To solve Equation (9.46) formally would require us to compute

$$\underline{u}_{n+1} = A^{-1}\left(B\underline{u}_n + \Omega_{n+1}\right). \quad (9.66)$$

As calculating the inverse of a matrix becomes infeasible for large values of M we have to rely on other methods of solution. Fortunately LAPACK offers several subroutines for doing such a task. Specifically, we have a symmetrical, tridiagonal matrix and thus require the subroutine 'DPTSV' to solve our system. Be aware that this subroutine performs an entire solve of the system by first factorising the matrix A, and using those factors to find the solution. In doing so it overwrites A with the factorised elements.

The program file *heatEqn1.f90* contains the code to perform the numerical solution of the one dimensional, time dependent heat equation for a specific set of boundary and initial conditions. As we have used the weighting parameter θ we can use this to set the finite difference method the code uses. Figure 9.6 shows some selected results of running the code with the ends of a one metre length metallic rod both held at 100°C, with an initial temperature of 0°C, and with $\theta = 0.5$, i.e. the Crank-Nicolson method. Here we set h=0.2 and k=0.01 making $r = 1/4$. We can see that the metallic rod behaves (at least qualitatively) as we might expect; heat flows into the rod from the hot ends in a smooth and symmetrical fashion, until the entire rod reaches 100°C after-which nothing happens i.e. it has reached a steady state. Our numerical results suggest this happens in around a second or less but, of course, we initially set the diffusion constant (related to the thermal conductivity of the metal) to one. It shouldn't be too much of a stretch to reintroduce it back into our program code. Figure 9.6 is somewhat of a cheat as it uses Excels' built in smoothing function to guess at the values in-between the grid points but you surely know how it does this by now, right? (If not, re-examine the chapter on interpolation).

The heat equation does have an analytical solution with Dirichlet boundary conditions that can be obtained through a technique called separations of variables. As we are treating the more general case of non-zero, Dirichlet boundary conditions this can be a little tricky. For a detailed derivation of this solution please visit http://tutorial.math.lamar.edu/Classes/DE/HeatEqnNonZero.aspx.

To recap we are solving the following PDE for the distribution of heat through a metallic rod of length L

$$U_t = kU_{xx}, \tag{9.67}$$

with boundary conditions

$$U(0,t) = c_1; \tag{9.68}$$

$$U(L,t) = c_2; \tag{9.69}$$

and with initial condition

$$U(x,0) = U_0(x). \tag{9.70}$$

The particular solution to this differential equation is given by

$$U(x,t) = U_E(x) + \sum_{n=1}^{\infty} D_n \sin\left(\frac{n\pi x}{L}\right) e^{-k\left(\frac{n\pi}{L}\right)^2 t}, \tag{9.71}$$

where the coefficients are

$$D_n = \frac{2}{L}\int_0^L (U_0(x) - U_E(x))\sin\left(\frac{n\pi x}{L}\right) dx, \tag{9.72}$$

and we have the steady state or equilibrium solution

$$U_E(x) = c_1 + \frac{c_2 - c_1}{L} x. \tag{9.73}$$

You should recognise this solution as being a Fourier series and is the solution (give or take some notation) Fourier found to the heat equation in his original works. We can see this solution makes physical sense because if $t \to \infty$ then the summation term in Equation (9.71) goes to zero and $U(x,t) \to U_E(x)$. Also if the initial conditions $U_0(x) = U_E(x)$, then the coefficients are zero for all n and we have the situation where there is no time dependence. This is of course what our physical intuition expects; if the temperature starts out at the steady state solution then it will remain at that solution for all time; that's why it's called a steady state solution.

We can use this analytical solution, Equation (9.71), to check the accuracy of our numerical approximation. I leave it as an exercise for you to modify the *heatEqn1.f90* to include this analytical solution to provide a measure of the error in the approximation. If we had no analytical solution how might we estimate the error in our approximation, or at least insure nominal significant figure accuracy?

9.4.2 The Heat Equation with Neumann Boundaries

Consider again the metallic rod but with one end insulated, rather than held at a constant temperature. In this case we are solving the system

$$U_t = kU_{xx}, \tag{9.74}$$

with the mixed boundary conditions of

$$U(0,t) = c; \tag{9.75}$$

$$U_x\big|_{x=L} = 0; \tag{9.76}$$

and with the initial condition

$$U(x,0) = U_0(x). \tag{9.77}$$

Equation (9.76) is telling us that there is no transfer of the physical quantity (temperature in this case) with respect to space across the specific boundary $x = L$. In other words the temperature gradient is zero across the boundary.

As the Neumann boundary condition, Equation (9.76), is a differential equation we need to replace it with a finite difference approximation. Recall that in order to maintain an error behaviour of the entire system our treatment at the boundaries of the domain should match that which we use on the interior of the domain. Fortunately, each of the methods discussed (implicit, explicit, and Crank-Nicolson) have an $\mathcal{O}(h^2)$ dependence on the space variable. Hence, any difference formula we use with an $\mathcal{O}(h^2)$ error behaviour to deal with the Neumann boundary condition will also apply to our general, weighted formula (Equation (9.44)).

One of the simplest ways of approximating Equation (9.76) with a finite difference equation is to use an $\mathcal{O}(h^2)$ forward or backward (depending on the location of the boundary) difference formula. The reason why we don't use a central difference formula is that it introduces a grid point that lays outside our computational domain, and while there are techniques to deal with this complication they are beyond the scope of our discussion here.

For our Neumann boundary we use the approximation

$$U_x\big|_{x=L} \approx \frac{1}{2h}\left(u_{M-2}^n - 4u_{M-1}^n + 3u_M^n\right) = 0, \tag{9.78}$$

for all n. This has an $\mathcal{O}(h^2)$ error behaviour. Rewriting Equation (9.78) we obtain the following expression for the boundary value at $x = L$

$$u_{M,n} = \frac{4}{3}u_{M-1}^n - \frac{1}{3}u_{M-2}^n. \tag{9.79}$$

We now have a means of calculating the value of U at the insulated boundary from the neighbouring interior grid points for a particular time step. But how does this affect our tridiagonal system of equations?

To answer that question we substitute Equation (9.79) into Equation (9.58) taken at the boundary $x = L$ such that

$$-(r\theta)\left(\frac{4}{3}u_{M-1}^{n+1} - \frac{1}{3}u_{M-2}^{n+1}\right) + (1 + 2r\theta)u_{M-1}^{n+1} - (r\theta)u_{M-2}^{n+1}$$
$$= \ldots \tag{9.80}$$

where ... represents the right hand side of Equation (9.58). We don't apply the substitution on the right hand side as we either know value of U at the boundary due to the initial conditions, or we calculate it from Equation (9.79). After some rearrangement we find the following

$$\left(1 + \frac{2}{3}r\theta\right)u_{M-1}^{n+1} - \left(\frac{2}{3}r\theta\right)u_{M-2}^{n+1} = \ldots \tag{9.81}$$

Notice that we haven't taken any terms over to the right hand side of these equations to deal with the Neumann boundary. Thus we make the following changes to our system of matrices and vectors. Matrix A remains unchanged except from its final two elements that now have the values $a_{M-2,M-3} = -2r\theta/3$, and $a_{M-2,M-2} = 1 + 2r\theta/3$. The additional boundary vector, Ω, is modified such that its final entry is now zero. Incorporating this change into the matrix B we see that its final element returns to being $b_{M-2,M} = r(1-\theta)$; remember that in the previous section we used the constant Dirichlet boundary conditions to simplify the first and last entries of B both to r. The vectors $\underline{u}$ and $\underline{\tilde{u}}$ need no modification.

The first thing to notice is that A is *no longer symmetrical*. This means we require a different LAPACK subroutine to solve for this system of equations. The subroutine 'DGTSV' solves a *general* tridiagonal system of linear equations using Gaussian elimination. As with the previous LAPACK subroutine the arrays representing A that are passed to 'DGTSV' are overwritten by their factors, hence they require resetting before looping to the next time step. The second thing to notice is that after we have solved the system at a particular time step we need to calculate the temperature at the Neumann boundary using Equation (9.79).

Figure (9.7) shows some of the results of running the program after making the necessary changes to the code found in *heatEqn1.f90*. Here we maintain the values of h and k from the previous section and the constant Dirichlet boundary is set to 100°C. Convince yourself that these results make physical sense. You may find, as I did, that the temperature at the Neumann boundary is a small but significant negative value for the first few time steps. This is likely due to the intrinsic error introduced by the coarseness of the spatial grid points. But, of course, you can test this can't you?

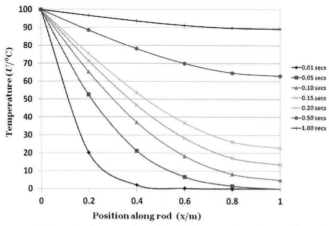

Figure 9.7: Numerical approximation to the transient flow of heat in a one dimensional metal rod with mixed boundary conditions

9.4.3 The Steady State Heat Equation

We have already established that the solution to the heat equation consists of a transient phase that evolves in time, which eventually settles on to a steady state solution. This steady state solution is also referred to as the equilibrium of the system. We can write a differential equation that expresses this steady state as

$$\nabla^2 U(x,y,z) \equiv \frac{\partial^2 U}{\partial x^2} + \frac{\partial^2 U}{\partial y^2} + \frac{\partial^2 U}{\partial z^2} = U_{xx} + U_{yy} + U_{zz} = 0, \quad (9.82)$$

where we have explicitly used our notation for partial differentiation. Here U is some physical field (temperature in this case) defined over three spatial dimensions; notice there is no time dependence. Equation (9.82) is called the *Laplace equation* and crops up often in physics wherever a system reaches an equilibrium state.

Consider a two dimensional metallic plate and we know the temperature of the plate at all points along its boundary (Dirichlet boundary conditions). We want to find the *steady state temperature* of the plate. This is similar to the one-dimensional metallic rod problem we discussed earlier in that we know the temperature at the boundaries of the system. However, in this steady state case we are not interested in the transient behaviour of the system. As such we are not required to know the initial state of the system in order to find a unique solution. In fact, we wish to solve this problem using an iterative or relaxation method.

In order to simplify the mathematics let's consider the metallic plate to be a rectangle of dimensions $a \times b$, with its lower left corner situated at the origin of our coordinate system. Thus we can write

$$U_{xx} + U_{yy} = 0, \quad (9.83)$$

with the following boundary conditions proceeding clockwise from the origin:

$U(0, y) = c_1;$

$U(x, b) = c_2;$

$$U(a, y) = c_3;$$

and

$$U(x,0) = c_4, \qquad (9.84)$$

for x on the interval $[0, a]$ and y on the interval $[0, b]$. For arguments sake the c_i are taken to be constants but in the general case they would be functions of x and y.

To solve this problem numerically we derive the finite difference approximation to the second ordered PDE of Equation (9.83) by defining a two dimensional grid on our rectangular plate. Hence we write

$$x_i = ih, \qquad i = 0,1,\ldots,N,$$

$$y_j = jk, \qquad j = 0,1,\ldots,M, \qquad (9.85)$$

where $h = a/N$ and $k = b/M$. The finite difference approximation can be written as

$$\frac{1}{h^2}\left(u_{i+1,j} - 2u_{i,j} + u_{i-1,j}\right) + \frac{1}{k^2}\left(u_{i,j+1} - 2u_{i,j} + u_{i,j-1}\right) = 0. \qquad (9.86)$$

Remember that we are going to solve this problem *indirectly* using a *relaxation method* so we need to solve Equation (9.86) for $u_{i,j}$. In doing so we find that

$$u_{i,j} = \frac{h^2 k^2}{2h^2 + 2k^2}\left(\frac{u_{i+1,j} + u_{i-1,j}}{h^2} + \frac{u_{i,j+1} + u_{i,j-1}}{k^2}\right). \qquad (9.87)$$

We can simplify Equation (9.87) further by imposing that $h = k$ and in this case

$$u_{i,j} = \frac{1}{4}\left(u_{i+1,j} + u_{i-1,j} + u_{i,j+1} + u_{i,j-1}\right). \qquad (9.88)$$

Note that this has a rather simple geometric interpretation. It states that the solution at a particular grid node is the arithmetic mean of its (four) nearest neighbours.

I leave it as an exercise for the reader to implement code to solve the steady state heat equation in two dimensions. Use the file *gauss-seidel.f90* as a guide; remember that you will need to express the array u in two dimensions, looping over both i and j, as well as initialising the values on the boundary and the initial guess for the interior grid nodes. As a comment on strategy, start out simple (e.g. small grid and use boundary conditions for which you can guess the solution) and add complexity once you have some working code. How might you visualise the data?

One such complication we might add is the inclusion of a Neumann boundary condition along one of the edges of our rectangular plate. For arguments sake let's replace the Dirichlet condition at the left hand edge with the Neumann condition that

$$U_x(x, y_j)\big|_{x=0} = \alpha_j \approx \frac{u_{1,j} - u_{-1,j}}{2h}, \qquad (9.89)$$

where we have introduced α_j as the partial derivative along the boundary $(0, y)$ for y on the interval $[0, b]$. Remember that the finite difference formula here is the first ordered central difference approximation and thus has $O(h^2)$ accuracy that matches the accuracy of the method used for the interior grid nodes. Using Equation (9.88) we can write the finite difference approximation for values along the left hand edge (i.e. $0 < j < M - 1$) as

$$u_{0,j} = \frac{1}{4}\left(u_{1,j} + u_{-1,j} + u_{0,j+1} + u_{0,j-1}\right). \qquad (9.90)$$

As we can see this equation involves a term that lies outside our physical grid. However, we can use Equation (9.89) to substitute in for $u_{-1,j}$, which gives

$$u_{0,j} = \frac{1}{4}\left(2u_{1,j} - 2h\alpha_j + u_{0,j+1} + u_{0,j-1}\right). \qquad (9.91)$$

In general α_j will be a function of x and y but is typically modelled by a constant value. The physics of Equation (9.89) is that heat will flow if there is a temperature gradient across a boundary, and that it will flow from hot to cold. For instance, if α was negative then heat flows into the plate from the environment, if α is positive that situation is reversed. Note that this direction of heat flow depends on the location of the boundary. For instance, if we consider the right hand edge of the plate then a negative temperature gradient (with respect to the x direction) implies heat flows *out-of* the plate into the environment, whereas a positive temperature gradient implies the opposite.

9.4.4 The Wave Equation

The general wave equation in one spatial dimension is given by

$$U_{tt}(x,t) = c^2 U_{xx}(x,t), \qquad (9.92)$$

where c is the speed of the wave, and U represents some physical field (for mechanical oscillations this will be the displacement of the wave). As with the heat equations studied previously to solve this equation numerically we define a discrete grid of points over the space and time dimensions, and then derive the finite difference approximation. Thus

$$\frac{1}{k^2}\left(u_m^{n+1} - 2u_m^n + u_m^{n-1}\right) = \frac{c^2}{h^2}\left(u_{m+1}^n - 2u_m^n + u_{m-1}^n\right), \qquad (9.93)$$

where we have imposed the discrete grid of points

$$\begin{aligned} x_m &= mh, & m &= 0,1,\ldots,M\,; \\ t_n &= nk, & n &= 0,1,\ldots,N\,, \end{aligned} \qquad (9.94)$$

with $h = (x_M - x_0)/M$ and $k = (t_N - t_0)/N$; we have yet to define the space and time boundaries of the system. For convenience we usually take $x_0 = 0$ and $t_0 = 0$.

As we are solving for a time dependent problem then we need to find an equation for the advanced time step in terms of values from previous time steps. Rearranging Equation (9.93) for the advanced time step we obtain

$$u_m^{n+1} = \rho\left(u_{m+1}^n + u_{m-1}^n\right) + 2(1-\rho)u_m^n - u_m^{n-1}, \quad (9.95)$$

where we have introduced

$$\rho = \left(\frac{kc}{h}\right)^2. \quad (9.96)$$

Interpreting Equation (9.95) we see that we can compute u for all x_m so long as we know u for all x_m at the previous two time steps. Note that this is an initial value problem in that to compute a solution we need to know the initial value of u for all x. But Equation (9.95) is telling us we need u at two previous time steps in order to advance the solution; we appear to be missing information. This dilemma is resolved when we realise that we have both the initial function u and can determine the first ordered time derivative of that initial function using a finite difference approximation. Explicitly

$$U_t(x_m,t)\Big|_{t=0} = \tau_m \approx \frac{u_m^1 - u_m^{-1}}{2k}, \quad (9.97)$$

where we have introduced τ_m for the partial derivative with respect to time for all x_m at the initial time $t = 0$. In order to calculate the first time step from the initial condition we use Equation (9.97) in Equation (9.95) to obtain the following expression

$$u_m^1 = \frac{\rho}{2}\left(u_{m+1}^0 + u_{m-1}^0\right) + (1-\rho)u_m^0 + k\tau_m. \quad (9.98)$$

Note that we used a finite difference approximation of τ in order to derive the formula for the first time step. However, τ might be some known function of x, for instance if the initial function $u(x,0)$ is easily differentiable, then we need not compute the finite difference approximation and merely compute $\tau_m = \tau(x_m)$ for each spatial grid point.

The astute reader will have already noticed that these conditions are essentially Cauchy boundary conditions; we can think of the initial time as being a boundary on the time dimension. You also may have noticed that we haven't yet imposed boundaries on the spatial coordinate. In fact we aren't required to impose such condition, however these leads to rather uninteresting and unphysical systems. Imagine, if you will, an infinitely long, stretched spring. We send a pulse wave down that spring by displacing it in some way; the tension in the spring provides the force to drive the pulse. Assuming no attenuation that wave continues to travel along the spring in the same direction for eternity. Boring, yes? But let me ask you this. How do you stretch a spring, and an infinite one at that? For normal, earthly, finite springs we have to fix at least one end (if not both) in order to stretch it and provide the tension required to carry a wave. This imposes the boundary conditions on our spatial coordinate.

It may or may not be of some surprise to you that it is in the modelling of those spatial boundary conditions rather than the wave equation itself that we find the most interesting physics. To illustrate, I sure you have all at some tedious moment, a boring lesson or lecture, or waiting for some experiment to finish, say, where you have twanged (technical term) your ruler on the side of whatever desk at which you've been sat. What are the spatial boundary conditions for that particular system? What if we were to fix both ends, or leave both ends free to oscillate? Would the outcome be as satisfying? For more serious musicians, those who play stringed instruments for example, it is more the interaction of the vibrating string with its support structures that is important for producing the instruments sound than is the vibrating string itself. Have you ever wondered why you don't see any square drums, or why brass instruments have a flared end?

Weird questions aside, write a program to implement the time dependent wave equation in one dimension, initially with fixed spatial boundaries. Here we don't need any fancy relaxation method or tridiagonal matrix solver as Equation (9.95) is explicit; we simply use that expression to advance our solution. The biggest issue here is in how you are going to visualise the data. You could save all the data to one large, two dimensional array, where the columns represent the time steps and the rows represent the spatial grid points, and then write that array to a text file, say, for inspection with a graphics program, such as gnuplot or Excel for example. Another, way would be have three, one dimensional arrays, one to hold all the u_m at time

t_{n-1}, a second to hold all the u_m at time t_n, and the third to hold all u_m at the advance time step, t_{n+1}. The arrays are shuffled accordingly. You could then save a snapshot of the system every z number of seconds, say, for later inspection with a graphics package. For the more ambitious out there you may want to find a way of animating your results instantaneously on screen (I recommend having a look at Java or Flash for this task).

9.5 Finite Element Method

It is at this point in the discussion of partial differential equations that most text books will mention the finite element method. To describe succinctly the method in easy to understand, plain English is a difficult thing to do especially at an introductory level to computational physics. Take note that the finite element method is extremely useful and can produce some highly accurate, and even beautiful solutions to some particularly nasty, and complex problems.

First lets expound the difference between the Finite Element Method (FEM) and the Finite Difference Method (FDM). In the FDM we take our computational domain and partition it into discrete points. The derivatives at those points are given by difference formulas, which we then use to approximate the governing differential equation. Along with the boundary and/or initial conditions, the resulting system of linear equations are solved. In this way we obtain an exact solution to an approximate problem. The FEM takes an alternative approach in that it uses a trial function, defined by some parameter, to estimate the solution and the resulting equations are solved in some best sense. In other words, it finds an approximate solution to the exact problem.

Next comes the precise formalism of the FEM. It involves talking about piecewise linear trial functions, basis functions, weighted residuals, and the Galerkin method. Rather than try to batter my way through these ideas I will point you in the direction of some decent texts on the subject (yes whole books are dedicated to the FEM).

Computational Differential Equations (1996) by K. Eriksson, D. Estep, P. Hansbo, and C. Johnson contains several chapters on the practical use of the FEM, as is accessible to the undergraduate student with some background knowledge (i.e. after having read this book) on numerical techniques.

An Introduction to the Finite Element Method (3rd ed.) by J. N. Reddy is more geared towards engineering undergraduates but does provide an excellent reference to the topic.

The Finite Element Method: A Practical Course (2003) by S. S. Quek and G.R. Liu provides an in-depth look at FEM and takes the reader through the basics of the method, providing examples and comprehensive discussions of applications and implementations.

The Finite Element Method: Volume 1: The Basis (5th ed.) by O. C. Zienkiewicz and R. C. Taylor provides a comprehensive and up-to-date overview of the topic and it accessible to undergraduate students. Volumes 2 and 3 are more complex but provide excellent grounding for anybody studying higher level FEM.

Plus lots more – too many to mention. I suggest looking through your universities resources for literature on FEM – there will be lots. See also the general reading guide on FEM at the back of this book.

Exercises

1. How might you go about setting up an actual experiment in the lab to test the accuracy of our numerical method for the heat distribution in a metallic rod?

2. The temperature of one end of a metallic rod is held at 0°C the other is held at 100°C. Using a finite difference method determine how the temperature of the rod evolves in time if the initial temperature of the rod was 20°C throughout. Ensure 4 significant figures of accuracy (Hint: use the analytical, Fourier series result, and/or Richardson extrapolation).

3. Consider the differential equation
$$y'' + 3y' - 5y = 7x$$
subject to the boundary conditions
$$y(0) = -20, \quad y(1) = 100.$$
Find a numerical solution to this equation using a direct finite difference method. Ensure at least 4 significant figures of accuracy.

4. Solve the same differential equation in the previous question but using a relaxation method. Ensure the same level of accuracy. Comment on any differences between the two methods.

5. The temperature of a unit square metal plate is subject to the following conditions
$$U(0, y) = e^{-10(y-0.5)^2}, \quad 0 < y < 1; \quad U(x,0) = U(x,1) = 100x$$
and the right hand side boundary is insulated. Find the steady state temperature of the plate.

6. A string on a guitar is plucked such that the initial function of the string can be described as
$$u(x,0) = \begin{cases} 0, & x = 0 \\ e^{-80(x-ct-0.5)^2}, & 0 < x < 1 \\ 0, & x = 1 \end{cases}$$
In other words, the string is of unit length, and held by rigid supports at its ends. The tension in the string is 12.8N and has a mass of 2g. Here the speed of the wave is $c^2 = T/\mu$ where T is the tension and μ is the mass per unit length. Evaluate what happens to the wave as time progresses. Did you observed phase reversal at the rigid supports?

7. Repeat the previous exercise but with one of the supports free. That is either
$$U_x\big|_{x=0} = 0 \text{ or } U_x\big|_{x=1} = 0.$$
(Hint: Use a finite difference approximation to find an expression for u_0^n or u_M^n)

Partial Differential Equations

8. A more realistic model is to assume the supports have inertial mass, M. If the (vertical) force on the supports is given by
$$F = TU_x\big|_{x=0,1}$$
find an expression for the boundary conditions and modify your program appropriately to study their behaviour in terms of their inertial mass.

9. Assuming the supports behave like damped, simple harmonic oscillators try to establish a more realistic model of the guitar string.

10 Advanced Numerical Quadrature

"The best way to become acquainted with a subject is to write a book about it."
— Benjamin Disraeli

The advanced nature that we boldly state in the chapter title is more to with the derivation of the methods rather than the application of the methods themselves. As with the majority of computational algorithms we don't need an in-depth understanding of why they work, just the knowledge that they do work and how to apply them. However, if we are going to use them we should make a little concerted effort to try to understand how they work in order to use them effectively.

This chapter covers the derivation and use of the Gauss-Legendre and Gauss-Laguerre quadrature. These two schemes will generate the most accurate numerical solutions for the least amount of computational effort, and should be used wherever possible in the numerical solution of a physical problem that involves integrals.

10.1 General Quadrature

In general, any integration can be approximated by a numerical quadrature written in the form of

$$\int_a^b f(x)dx \approx \sum_{m=1}^N w_m f(x_m), \qquad (10.1)$$

where x_m are the evaluation points, w_m are weights given to the mth point, and there are N evaluation points in total. For convenience the points are set with uniform spacing, typically denoted by h, and we can derive the numerical quadrature methods as discussed in Chapter 5 (e.g. trapezoidal rule, Simpson's rule etc.). To do this we begin by deriving the simplest formula from Equation (10.1) which is to only consider the limits of the integration a and b thus

$$I \approx w_1 f(x_1) + w_2 f(x_2) \tag{10.2}$$

where $x_1 = a$ and $x_2 = b$. In the limit of the integration interval $(b - a)$ going to zero we require that Equation (10.2) be exact for any function. This sounds like a difficult task, however if we consider any function that has a Taylor series expansion we note that the first two terms of that expansion are multiples of 1 and x (note that sometimes the multiple is zero c.f. sine). Hence, we now attempt to find the weights, w_1 and w_2, that will give

$$\int_{x_1}^{x_2} 1 dx = x_2 - x_1 = w_1 + w_2 \tag{10.3}$$

and

$$\int_{x_1}^{x_2} x dx = \frac{x_2^2 - x_1^2}{2} = w_1 x_1 + w_2 x_2 \tag{10.4}$$

Note the equivalencies; these equations are *exact*. We have two equations and two unknowns, namely the weights. Solving for the weights we find

$$w_1 = w_2 = \frac{x_2 - x_1}{2}. \tag{10.5}$$

Typically we would write $h = x_2 - x_1$ and our approximation for the integral of any function becomes

$$\int_a^b f(x) dx \approx \frac{h}{2}(f_1 + f_2), \tag{10.6}$$

which is the (primitive) trapezoidal rule. Note that it is from this derivation that we can explicitly state the error behaviour of a numerical quadrature using what is called the Lagrange remainder (this is derived using the mean value theorem applied to the remainder term in the Taylor series expansion)

$$\int_a^b f(x)dx = \frac{h}{2}(f_1 + f_2) - \frac{h^3}{12} f(c), \qquad (10.7)$$

where c lies somewhere in the integration interval.

Of course we can add more evaluation points. If we now take three points $x_1 = a$, $x_2 = (b+a)/2$, and $x_3 = b$ such that $h = (b-a)/2$ we can write the following equivalencies,

$$x_3 - x_1 = w_1 + w_2 + w_3, \qquad (10.8)$$

$$\frac{x_3^2 - x_1^2}{2} = w_1 x_1 + w_2 x_2 + w_3 x_3, \qquad (10.9)$$

and

$$\frac{x_3^3 - x_1^3}{3} = w_1 x_1^2 + w_2 x_2^2 + w_3 x_3^2, \qquad (10.10)$$

where we have use the first three terms of the Taylor series expansion. Here we have three equations and three unknowns, and solving for the weights gives us Simpson's rule

$$\int_{x_1}^{x_3} f(x)dx = \frac{h}{3}(f_1 + 4f_2 + f_3) - \frac{h^5}{90} f^{[4]}(c). \qquad (10.11)$$

Adding another evaluation point leads to Simpson's three-eighths rule, five points leads to Boole's rule, and so on. As a reminder all these formulas require that the $f(x)$ be expressible as a polynomial (i.e. have a Taylor series expansion). Indeed, usually these integration rules are derived by considering a polynomial approximation to the function and integrating that approximation exactly. For instance trapezoidal rule is a linear approximation, Simpson's rule is a quadratic approximation, and so forth.

In the treatment above we constrained the evaluation points to be equally spaced. However, this is not a requirement and we can take the evaluation points to be anywhere within the integration region. In fact, by removing this constraint we can derive more accurate numerical quadrature methods, but they are not so easy to derive as we increase the number of evaluation points.

To demonstrate, if we consider the simplest integration formula in which a single evaluation point that can be located anywhere within the integration interval then we now have two unknowns, namely the weight *and* the evaluation point. To solve for these two unknowns we need two equations and as before we obtain these equations by requiring that the quadrature be exact for the first two lowest ordered polynomials $f(x) = 1$, and $f(x) = x$. This gives the necessary equations thus

$$(b-a) = w_1 \tag{10.12}$$

and

$$\frac{(b^2 - a^2)}{2} = w_1 x_1. \tag{10.13}$$

Solving for both the evaluation point and the weight we find that $x_1 = (b + a)/2$ and $w_1 = b - a$. This is the mid-ordinate rule and is exact for linear functions. Note that the trapezoidal rule is also exact for linear functions but we have to take an extra function evaluation. In fact the mid-ordinate rule can be considered as the first ordered Gauss-Legendre quadrature, though normally taken on the normalised integration interval $[-1,1]$; more on this shortly.

If we now add a second evaluation point we now have to find four unknowns which requires four equations. Again we require that the quadrature be exact for the first four lowest polynomials such that

$$(b - a) = w_1 + w_2, \tag{10.14}$$

$$\frac{1}{2}(b^2 - a^2) = w_1 x_1 + w_2 x_2, \tag{10.15}$$

$$\frac{1}{3}(b^3 - a^3) = w_1 x_1^2 + w_2 x_2^2, \tag{10.16}$$

and

$$\frac{1}{4}(b^4 - a^4) = w_1 x_1^3 + w_2 x_2^3. \qquad (10.17)$$

In order to solve these equations for the weights and abscissas we have to normalise the integration region such that the interval $[a, b]$ is mapped to the interval $[-1, 1]$. After the equations have been rewritten in terms of this mapping it is a (relatively) straight forward task of finding the weights and abscissas. After performing the necessary steps we find that the weights are equivalent and equal to one and, the evaluation points, or abscissas as they are technically known, are $x_1 = \sqrt{(1/3)}$ and $x_2 = -\sqrt{(1/3)}$. Note that the quadrature

$$\int_{-1}^{1} f(x)dx \approx f\left(\frac{1}{\sqrt{3}}\right) + f\left(-\frac{1}{\sqrt{3}}\right) \qquad (10.18)$$

is now exact if $f(x)$ is a polynomial of order three or less! Remember the trapezium rule is only exact for linear functions (order one or zero). This is the seconded order Gauss-Legendre quadrature.

If we now try three points we find that we have six unknowns and thus require six equations. If you've spotted the pattern then we require the quadrature to be exact for the first six lowest ordered polynomials, i.e. up to $f(x) = x^5$. Extending this to N points, we have $2N$ unknowns and therefore require $2N$ equations. This means an N point Gauss-Legendre quadrature is exact for polynomials of order $2N - 1$ and less.

The job then is to find the weights and corresponding abscissas for each of those N points. Rather than trying to continue to solve sets of simultaneous equations, which would get somewhat difficult (and tedious), let's turn to another method; orthogonal polynomials[*]. It is from these that Legendre gets his name appended to the method (Laguerre also).

---

[*] Collective groan from the audience.

10.2 Orthogonal Polynomials

You may have heard the term orthogonal banded about the place when discussing vectors, or considering Cartesian coordinates. For instance, to determine whether or not two vectors are orthogonal we take what is known as their scalar product. This product is also known as the dot product or the inner product. If the resulting outcome of the inner product is zero we know that the two vectors are orthogonal. Essentially, orthogonal is another word for perpendicular or at right-angles to but has a deeper meaning when applied to functions; it means they are fundamentally different.

Orthogonal polynomials are a set of polynomials φ_m defined over a finite range $[a, b]$ such that they obey an orthogonality relation given by

$$\int_a^b w(x)\varphi_m(x)\varphi_n(x)dx = \delta_{mn}c_n \qquad (10.19)$$

where $w(x)$ is a weighting *function*, δ_{mn} is called the Kronecker delta that is equal to 1 when $m = n$ and zero otherwise, and c_n is some constant coefficient.

It would be a hopeless task to try to identify a set of orthogonal polynomials via substitution into Equation (10.19) however that is not the purpose of the relation. We'll do the opposite and use the relation to *construct* a set of orthogonal polynomials. To illustrate this process, let's make life easier for ourselves and reduce the complexity of the relation somewhat. Let's assume we are on the normalised integration interval such that $a = -1$ and $b = 1$, and let's also assume our weighting function is constant and equal to one. By choosing the integration limits as such we are not losing any generality as the interval can always be mapped back on to any finite region through a change of variables. With these simplifications in place we can construct the first polynomial, φ_0, using

$$\int_{-1}^{1} \varphi_0(x)\varphi_0(x)dx = c_0 . \qquad (10.20)$$

It is often convenient to normalise the polynomials such that all $c_n = 1$, in which case the polynomials are referred to as orthonormal (a contraction of orthogonal and normalised)*. There are many polynomials that satisfy Equation (10.20) so let's choose the simplest (non-trivial) case that is $\varphi_0 = k_0$, where k_0 is constant. Performing the integration we find that $2k_0^2 = 1$, hence

$$\varphi_0 = 1/\sqrt{2}. \tag{10.21}$$

We find the next polynomial φ_1 in the set by requiring that

$$\int_{-1}^{1} \varphi_0(x)\varphi_1(x)dx = 0 \tag{10.22}$$

It is tempting here to just set φ_1 equal to x, which would definitely satisfy Equation (10.21). However we should be more general in our approach. Our strategy is to assume that the Nth ordered polynomial of the orthogonal set is given by the linear combination

$$\varphi_N(x) = k_N \left[u_N(x) + \alpha_{NN-1}\varphi_{N-1}(x) + \ldots + \alpha_{N0}\varphi_0(x) \right] \tag{10.23}$$

where $u_m = x^m$, and we can use the k_m to normalise φ_m, and the α_{mn} are chosen to force orthogonality. We have already found the first polynomial of the set, $\varphi_0 = 1/\sqrt{2}$, hence we write

$$\varphi_1(x) = k_1 \left(x + \frac{\alpha_{10}}{\sqrt{2}} \right) \tag{10.24}$$

and we can force the integral of Equation (10.21) to be zero by choosing an appropriate α_{10}. Substituting Equation (10.23) into (10.21) we find that α_{10} is zero. The normalisation constant, k_1, is found by performing

$$\int_{-1}^{1} \varphi_1(x)\varphi_1(x)dx = k_1^2 \frac{2}{3} = 1, \tag{10.25}$$

such that the next orthonormal polynomial in the set is

---

* c.f. chillax.

$$\varphi_1(x) = \sqrt{\frac{3}{2}} x. \qquad (10.26)$$

The next polynomial in the set is then found by the linear combination

$$\varphi_2(x) = k_2\left(x^2 + \alpha_{21}\varphi_1(x) + \alpha_{20}\varphi_0(x)\right). \qquad (10.27)$$

Now, we require that φ_2 is orthogonal to both φ_1 and φ_0 such that

$$\int_{-1}^{1} \varphi_0(x)\varphi_2(x)dx = 0 \qquad (10.28)$$

and

$$\int_{-1}^{1} \varphi_1(x)\varphi_2(x)dx = 0. \qquad (10.29)$$

After performing the necessary calculations we find that $\alpha_{21} = 0$ and $\alpha_{20} = -\sqrt{2}/3$. Again we normalise the polynomial by finding k_2 using

$$\int_{-1}^{1} \varphi_2(x)\varphi_2(x)dx = 1. \qquad (10.30)$$

After some manipulation we find that

$$\varphi_2 = \sqrt{\frac{5}{2}} \frac{3x^2 - 1}{2}. \qquad (10.31)$$

We can continue in this fashion to find any order of polynomial that fits in this orthonormal set. It is of consequence that Equations (10.21), (10.26), and (10.31) are the first three (normalised) Legendre polynomials.

The Legendre polynomials are the orthogonal set specific to the weighting function equal to one and for the integration range [-1,1].

For other weighting functions and integration limits we would necessarily construct a different set of orthogonal. For instance with $w(x) = e^{-x}$ on the integration range $[0, \infty]$ we would arrive at the Laguerre polynomials.

The general process of finding a set of orthogonal (orthonormal) polynomials in this way is due to Jorgen Pedersen Gram, a Danish mathematician, and Erhard Schmidt, a German mathematician who developed the eponymous Gram-Schmidt process in the late 19th century.

10.3 Gauss-Legendre Quadrature

To apply our new found knowledge of orthogonal polynomials to the Gauss quadrature we reformulate our integration such that

$$\int_a^b f(x)w(x) = \sum_{m=1}^N w_m f(x_m) \qquad (10.32)$$

where the weighting function $w(x)$ is what is known as positive definite, i.e. it is never negative, and is the same function as used with the orthogonal polynomials. We have $2N$ unknowns in this equation due to the weights and abscissas. Thus we require that integration of the polynomials of order up to and including $2N - 1$ are to be given exactly by the quadrature, and we use this requirement to define the weights and abscissas for the quadrature.

To do this we let $f(x)$ be some arbitrary polynomial of order $2N - 1$, and we define φ_N as an orthogonal polynomial of order N that is particular to the weighting function $w(x)$ and the region of the integration $[a, b]$ expressed in Equation (10.32). If we now divide $f(x)$ by φ_N we obtain a quotient term, q, and a remainder term, r, both of which will be polynomials of order $N - 1$. Remember this is polynomial long division, something you should have covered in an A-level (or equivalent) mathematics course; I suggest you dig your notes out if you didn't burn them.

The integral of Equation (10.32) can now be express as

$$\int_a^b f(x)w(x) = \int_a^b q_{N-1}(x)\varphi_N(x)w(x)dx + \int_a^b r_{N-1}(x)w(x)dx. \qquad (10.33)$$

For the remainder of this argument we can ignore the remainder term (insert groans here) as it is no longer required in our derivation of the weights and abscissas for the quadrature. The quotient polynomial can be expanded as a linear combination of a set of polynomials ranging in order from zero to $N-1$. Fortunately we already have a (complete) set of polynomials that we can use for this task; $\{\varphi_m\}$, the orthogonal set of polynomials. (The curly braces indicate a set, and the φ_m are members of the set.) Explicitly we write

$$q_{N-1}(x) = \sum_{m=0}^{N-1} d_m \varphi_m(x), \qquad (10.34)$$

where the d_m are constants. With this expansion we can now express the integral of the quotient term as

$$\int_a^b q_{N-1}(x)\varphi_N(x)w(x)dx = \sum_{m=0}^{N-1} d_m \int_a^b \varphi_m(x)\varphi_N(x)w(x)dx$$
$$= \sum_{m=0}^{N-1} d_m \delta_{mN} c_N = 0 \qquad (10.35)$$

The zero emerges as the summation only goes up to $N-1$ and for the Kronecker delta to be non-zero (i.e. one) m has to equal N.

At the start of our argument we required that polynomials of order $2N-1$ are given exactly by the quadrature. The product $q_{N-1}(x)\varphi_N(x)$ is a polynomial of order $2N-1$ therefore we can write

$$\int_a^b q_{N-1}(x)\varphi_N(x)w(x)dx = \sum_{m=1}^{N} w_m q_{N-1}(x_m)\varphi_N(x_m) = 0. \qquad (10.36)$$

As we have kept the argument general q_{N-1} is an arbitrary polynomial and as such is not necessarily zero at the abscissas. The only way to ensure the sum is zero is to require that all the $\varphi_N(x_m)$ are zero, ignoring the trivial case where all the weights are zero. In other words, we find the roots of the polynomial $\varphi_N(x)$ and select them as our abscissas. As an N order polynomial will have N roots (ignoring the case were the roots are complex) we have found all the abscissas for our N point Gauss quadrature. Now for the weights.

As the quadrature is exact for polynomials of order $2N - 1$ it must also be exact for polynomials of lesser order. Here we have the freedom to choose any polynomial that has an order less than $2N - 1$ however, we should choose one that simplifies the mathematics. Fortunately others that have come before us have identified the polynomial we need. Lagrange's* interpolating polynomial, or more specifically, the multiplication factor (Equation (3.14)) has the properties we desire. Rewriting it here in terms of our current parameters

$$\lambda_{i,N}(x) = \frac{\prod_{l=1 \neq i}^{N}(x - x_l)}{\prod_{l=1 \neq i}^{N}(x_i - x_l)} \tag{10.37}$$

where the x_i and x_l are the abscissas. This polynomial is of order $N - 1$ and has the property that

$$\lambda_{i,N}(x_m) = \begin{cases} 0, m \neq i \\ 1, m = i. \end{cases} \tag{10.38}$$

We are therefore able to write that

$$\int_a^b \lambda_{i,N}(x) w(x) dx = \sum_{m=1}^{N} w_m \lambda_{i,N}(x_m) = w_i. \tag{10.39}$$

In other words we can find the weights of the corresponding abscissas by performing the analytical integration on the left-hand-side of Equation (10.39).

Currently, this treatment has been general in that we have not defined our integration limits or the weighting function. Let us do

---

* What is it with all these mathematician's names starting with 'L'?

this now. We know from our discussion on orthogonal polynomials that by setting our integration region to $[-1,1]$ and choosing $w(x) = 1$, we obtain the Legendre polynomials as our orthogonal set. Hence the abscissas for the Gauss-Legendre quadrature are the roots of the Legendre polynomials. Once we have found those roots we use Equation (10.39) with the appropriate parameters to compute the weights.

To illustrate, consider Gauss-Legendre quadrature with two points. We use the (normalised) Legendre polynomial

$$\varphi_2 = \sqrt{\frac{5}{2}} \frac{3x^2 - 1}{2}, \tag{10.40}$$

which has roots

$$x_m = \pm \frac{1}{\sqrt{3}}. \tag{10.41}$$

Note that we'd obtain the same roots using the non-normalised polynomial.

Performing the integration of Equation (10.39) with the appropriate parameters explicitly gives

$$w_1 = \int_{-1}^{1} \lambda_{1,2} dx = \int_{-1}^{1} \frac{x - x_2}{x_1 - x_2} dx = \frac{1}{x_1 - x_2} \left[\frac{x^2}{2} - x_2 x \right]_{-1}^{1} = \frac{-2x_2}{x_1 - x_2} = 1$$

and

$$w_2 = \int_{-1}^{1} \lambda_{2,2} dx = \int_{-1}^{1} \frac{x - x_1}{x_2 - x_1} dx = \frac{1}{x_2 - x_1} \left[\frac{x^2}{2} - x_1 x \right]_{-1}^{1} = \frac{-2x_1}{x_2 - x_1} = 1.$$

This is exactly the same result we got before by solving the set of (non-linear) simultaneous equations, Equations (10.14-17). Although getting here was tough I hope you'll agree that finding the abscissas and weights for $N = 2$ using orthogonal polynomials was easier than solving a set of non-linear simultaneous equations. Even you don't we now have a general method to obtain the weights and abscissas for any number points N, which was worth it.

In fact you don't actually have to do this work as the quadrature is so frequently used that the weights and corresponding abscissas have been published and tabulated, several times over, and to varying degrees of precision. Just type 'Gauss-Legendre weights and abscissas' into your favourite search engine and you'll find them.

10.4 Programming Gauss-Legendre

There are two ways forward to programming the Gauss-Legendre quadrature. We can either enter all the weights and abscissas into a PARAMETER expression in a module and include that module whenever we write a program that involves the quadrature. Or we can write a subroutine or function that evaluates the weights and abscissas for us every time we wish to perform the quadrature. Both methods have their advantages. Storing the values is a good idea for speed as the values need only be read from memory to be used, though they would have to be entered with care; a typo in a list of numbers can be very tedious to track down. In addition to this, you are limited to the number of weights and abscissas you can be bothered to enter as well as their precision. Whereas the subroutine or function method gives you the flexibility to decide to use, say, a 20 point Gauss-Legendre quadrature if you so wished, and to quadruple precision. However, this means these values have to be computed every time you perform an integration which may add valuable time on to your overall computation.

Instead of giving you the tabulated numbers to some precision or other (which you can find for yourselves with only a marginal amount of effort) let's do the subroutine/function method. In this way if you are concerned with performance, you could actually store the values obtained to a text file, say, and then read those values from the file when required.

Advanced Numerical Quadrature

Legendre polynomials are defined by the following recursive rule (note that we've dropped the normalisation[*])

$$\varphi_0 = 1 \tag{10.42}$$

$$\varphi_1 = x \tag{10.43}$$

$$\varphi_n(x) = \frac{1}{n}\left[(2n-1)x\varphi_{n-1}(x) - (n-1)\varphi_{n-2}(x)\right]. \tag{10.44}$$

The roots of φ_n are not generally analytically soluble so we have to apply a root finding algorithm. Our Newton-Raphson algorithm will perform the job nicely. We can use Newton-Raphson rather than the secant method as we can determine the analytical first ordered derivative of the φ_n from Equation (10.44). Explicitly the recursion relation for the derivatives are

$$\varphi'_n(x) = \frac{n}{x^2 - 1}\left(x\varphi_n(x) - \varphi_{n-1}(x)\right). \tag{10.45}$$

To speed up our root searches we use the fact that the first guess x_0 for the ith root of a n-order polynomial φ_n can be given by

$$x_0 = \cos\left(\pi \frac{i - 1/4}{n + 1/2}\right). \tag{10.46}$$

As Equation (10.46) gives us a relatively decent estimate of the root we don't need the robustness of a bisection method in our root search. After we get the abscissas x_m via the root search to some precision we compute the appropriate weights by

$$w_m = \frac{2}{(1 - x_m^2)[\varphi'_n(x_m)]^2}. \tag{10.47}$$

Once the weights and abscissas are computed for a N point quadrature, we can approximate an integral over any interval $[a, b]$ by

---

[*] Just 'cos, that's why!!

$$\int_a^b f(x)dx \approx \frac{b-a}{2}\sum_{m=1}^N w_m f\left(\frac{b-a}{2}x_m + \frac{a+b}{2}\right). \tag{10.48}$$

We have provided the file *GaussLeg.f90* that contains the code to perform the Gauss-Legendre quadrature. Here we use some new syntax which shows off the more modern features (and power) of the Fortran language. At the top of the file we set an integer 'p' that will define the precision of the values declared using the 'kind' attribute. With p = 16 we get quadruple precision! You may have already guess that the value defines the number of bytes to use for the variable with single precision given by 4 and double precision given by 8. When assigning values to variables we use '_p' to denote the precision we want. Note that I have left the p as a lower case to distinguish it from program variables. The array *R* stores our abscissas and corresponding weights with its dimension being defined by the function 'GUASSQUAD'.

The function itself computes the Legendre polynomial *coefficients*, storing them to array *P1*; *P0* and *TMP* are required to facilitate the computation. Note that these arrays are special in that they expand as we cycle through the loop. The square brackets essentially concatenate the elements placed inside them. For instance, for the first loop the term [*P1*, 0._p] is actually an array of length 3 as *P1* is an array of length 2 as we enter the loop. You may have already seen this construct if you've done some MATLAB or Octave programming.

Once we have the coefficients of the Nth Legendre polynomial we then find its roots. As the cosine function gives a relatively good estimate of the root, the Newton-Raphson method will always converge to the required root of the polynomial. It is not obvious that the code here calculates the polynomial function and its derivative for the given x_m. It uses the array of coefficients, *P1*, to build up the function and its derivative for use with the Newton-Raphson root search. You may find it instructive to perform the calculations manually for small *N* or have the program print out the polynomial values as a check. Once the roots are found to the desired precision or we have performed ten iterations of the root search, we store them to the first row of the result array R. The corresponding weights are calculated and stored to the second row of R.

The result array is passed back to the main program where we compute the actual integral using the Fortran intrinsic 'DOT_PRODUCT' function. This saves us having to write out the do loop for the summation found in Equation (10.48). In this case we are performing the example integration

$$\int_1^4 e^x dx = e^4 - e^1 = 51.8798682.... \qquad (10.49)$$

The file prints the number of points used in the quadrature, the value obtained, and the error in the solution.

After you compile and run this program you should find that the error reduces rather rapidly as we increase the number of points used. For 14 points we obtain a value that has an error on the order of 10^{-30}, this means the value printed is accurate up to at least 29 significant figures, if not 30. This is the power of the Gauss-Legendre quadrature. However, any more than 14 points and we see that the error begins to creep back up again. Thus taking more points does not necessarily guarantee more accuracy. We can circumvent this by using a composite strategy instead. As with the other integration rules an increase in accuracy was obtained by segmenting the integration interval into smaller strips, and performing a relatively simple quadrature on each strip. Try this now for the Gauss-Legendre quadrature and see if you can improve on the accuracy in any way.

10.5 Gauss-Laguerre Quadrature

One limitation to the Gauss-Legendre quadrature is that it only applies to integrals with finite limits. In physics it often happens that a physically significant quantity can be given by the semi-infinite integral

$$I = \int_0^\infty g(x)dx. \qquad (10.50)$$

For the integral to be finite $g(x)$ must vanish more rapidly than the inverse of x (c.f. convergence of an infinite sum). One way to ensure this condition is to recast where possible the integrand function as

$$I = \int_0^\infty g(x)dx = \int_0^\infty e^{-x} f(x)dx, \qquad (10.51)$$

as the exponential weight definitely vanishes more quickly than $1/x$. We now have an integral in the form of the left hand side of Equation (10.32). As before, we need to find a set of orthogonal polynomials that will satisfy these particular limits and the weighting function.

Fortunately this work has already been done before and the set of polynomials we need are the Laguerre polynomials. We can define the Laguerre polynomials recursively, defining the first two polynomials as

$$\varphi_0(x) = 1, \qquad (10.52)$$

and

$$\varphi_1(x) = 1 - x, \qquad (10.53)$$

then using the following recurrence relation for any $n \geq 1$:

$$\varphi_{n+1}(x) = \frac{1}{n+1}\left[(2n+1-x)\varphi_n(x) - n\varphi_{n-1}(x)\right]. \qquad (10.54)$$

Following the same strategy as before we find the roots of the Laguerre polynomial. To do this we note that

$$\varphi'_n(x) = \frac{n}{x}\left(\varphi_n(x) - \varphi_{n-1}(x)\right). \qquad (10.55)$$

Unfortunately, there isn't a nice formula for the estimation of the roots so you may find useful to plot the Laguerre polynomials and find estimations for the roots manually. A numerical recipes handbook may offer more guidance here.

Once the roots are found they are used to obtain the weights using the relation

$$w_m = \frac{x_m}{(n+1)^2 [\varphi_{n+1}(x_m)]^2}. \tag{10.55}$$

Note that the denominator contains a factor of the polynomial squared evaluated at the abscissa *not* the derivative as with the Legendre weights.

Of course, the weights and abscissas have been tabulated and published elsewhere. Should wish just use them rather than calculate them then you should be able to find them in the literature.

As the upper limit of the integral is infinity we cannot derive a composite formula for the quadrature. If higher accuracy is needed than the Gauss-Laguerre quadrature can achieve, you can always separate the semi-infinite region into two, using a composite Gauss-Legendre quadrature up to some finite limit, and then a Gauss-Laguerre quadrature from that limit up to infinity. Though, remember to adjust the variables for the change in the lower limit of the integration.

You may have already guessed that their exists several other Gaussian quadrature methods for different integration limits and weighting functions. Of note are the Gauss-Hermite, Gauss-Chebyshev, and Gauss-Jacobi quadrature methods, which you should look up. Though different they all share the same common algorithm – define the set of polynomials to use, find the roots of those polynomials, use those roots to compute the corresponding weights, and finally compute the quadrature for a given number of points.

Exercises

1. Either using the file provided or with your own program compute

$$\int_{-1}^{1} x^m \, dx$$

for $m = 0, 1, \ldots, 15$ using Gauss–Legendre quadrature to double precision. If the code is correct how accurate should the quadrature be for the appropriate number of points used?

2. Compare the effort required to compute

$$\int_{\mu-\sigma}^{\mu+\sigma} \frac{1}{\sigma\sqrt{2\pi}} e^{-\frac{(x-\mu)^2}{2\sigma^2}} \, dx$$

to 10 significant figures for the Trapezoidal rule, Simpson's rule, and Gauss-Legendre quadrature. You may find it illustrative to plot the relative error against the number of points used for each method.

3. Write a program to compute the weights and abscissas for the Gauss-Laguerre quadrature. Use the file *GaussLeg.f90* as a guide. To test your program evaluate the integral

$$\int_{0}^{\infty} e^{-x} \, dx \, ;$$

it should be simply the sum of the weights for the number of points used.

4. In Planck's treatment of black body radiation the following integral appears:

$$\int_{0}^{\infty} \frac{x^3}{e^x - 1} \, dx$$

Evaluate the integral ensuring 10 significant figures of accuracy.

5. The integral

$$f(\theta) \approx -\frac{2m}{\hbar^2}\int_0^\infty rV(r)\sin(2kr\sin(\theta/2))dr$$

appears in the theory for the cross section of a quantum scattering event. It describes the force felt by a quantum particle as it interacts with the scattering potential V as a function of the incident angle θ. If the particle is an electron scattering from an atomic nucleus then the scattering potential is given by

$$V(r) = \frac{1}{4\pi\varepsilon_0}\frac{Zq^2}{r}e^{-r/r_0}$$

where Z is the proton number of the nucleus, q is the proton charge and r_0 is the so-called screening length. Plot f as a function of θ for an atom of your choice; use the Bohr radius as the screening length.

11 Advanced ODE Solver and Applications

"Either do, or do not. There is no try."

- Yoda[*]

In this chapter we explore a more advanced ODE solver and how we can apply it to solve some "difficult" physics problems from finding chaos in a driven pendulum to sending a spaceship to the Moon (and beyond) to the wavefunctions of electrons in an arbitrary electrical potential. We will also show how to use the solver in combination with the Fast Fourier Transform subroutine in Chapter 7 to analyse the frequency spectrum of the Van der Pol oscillator, which will provide the guide for you to analyse the spectrum of the chaotic pendulum.

Of course as most of these systems are dynamical it would be nice to be able to animate them in some way using Java or Flash animation for example...but that is a task for another day (or perhaps a challenge to the reader).

11.1 Runge-Kutta-Fehlberg

In Chapter 6 we explored the use of the finite difference method to solve ODEs. During this discussion we developed a technique to make the step sizes adapt to the local nature of the solution by halving the step size when it was too large and generating too much local error, or doubling the step size when it was too small and wasting computational effort. Although this method is effective there is a better way of making the step size adaptive, rather than just halving or doubling its length.

Erwin Fehlberg published a number of adaptive step Runge-Kutta methods in two NASA technical reports in 1968 and 1969. In Fehlberg's algorithm, two Runge-Kutta methods of different order

---

[*] George Lucas really.

are run simultaneously. At each step, the lower ordered method is computed first, producing an estimate of the solution, y_{n+1}. Next, the second, higher ordered method is computed with more function evaluations producing the estimate $\hat{y}_{n+1}$ for the same step size. The difference between these two methods gives an estimate of the local error for the step size used. If this estimation of the error is within the prescribed tolerance the step is accepted and solution is advanced. If not the step is rejected and the process is repeated with a reduced step size. Whenever a step is accepted the next step size is estimated using the values obtained from y and $\hat{y}$, and we use the more accurate value of $\hat{y}$ as our initial value for the next step; more on this shortly.

At first glance, this may not seem like we're saving much on computational resource; although the steps will be adaptive we have to make several extra function evaluations at each step in order to compute the two Runge-Kutta methods. However, the beauty of Fehlberg's algorithms is that he found coefficients such that the two methods share function evaluations and only a few extra evaluations are required for the higher ordered method. One such Fehlberg algorithm is based on the classic fourth ordered Runge-Kutta, with a fifth ordered Runge-Kutta used as the higher ordered estimate. The informed reader may now be guessing that this is why one of the differential equation solvers in MATLAB/OCTAVE is named 'ode45'.

In the Runge-Kutta-Fehlberg fourth-fifth (RKF45) algorithm each accepted step requires a total of six intermediary function evaluations; four for the 4th order Runge-Kutta, and two more for the 5th order Runge-Kutta. Fehlberg's equations for the RKF45 method are as follows: the intermediary function evaluations are given by

$$k_0 = f(t_0, y_0), \tag{11.1}$$

$$k_1 = f\left(t_0 + \frac{h}{4}, y_0 + \frac{h}{4}k_0\right), \tag{11.2}$$

$$k_2 = f\left(t_0 + \frac{3h}{8}, y_0 + \frac{3h}{32}k_0 + \frac{9h}{32}k_1\right), \tag{11.3}$$

$$k_3 = f\left(t_0 + \frac{12h}{13}, y_0 + \frac{1932h}{2197}k_0 - \frac{7200h}{2197}k_1 + \frac{7296}{2197}k_2\right), \quad (11.4)$$

$$k_4 = f\left(t_0 + h, y_0 + \frac{439h}{216}k_0 - 8hk_1 + \frac{3680h}{513}k_2 - \frac{845h}{4104}k_3\right), \quad (11.5)$$

$$k_5 = f\left(t_0 + \frac{h}{2}, y_0 - \frac{8h}{27}k_0 + 2hk_1 - \frac{3544h}{2565}k_2 + \right.$$

$$\left. \frac{1859h}{4104}k_3 - \frac{11h}{40}k_4\right), \quad (11.6)$$

where f is the function defining the differential equation, h is the step size, t is the independent variable, and y is the dependent function. The index zero refers to the values at the beginning of a given step. Using these definitions for the intermediary function evaluations the fourth ordered Runge-Kutta is written as

$$y_{n+1} = \hat{y}_n + h\left(\frac{25}{216}k_0 + \frac{1408}{2565}k_1 + \frac{2197}{4104}k_3 - \frac{1}{5}k_4\right) \quad (11.7)$$

and the fifth ordered Runge-Kutta is written as

$$\hat{y}_{n+1} = \hat{y}_n + h\left(\frac{16}{135}k_0 + \frac{6656}{12825}k_1 + \frac{28561}{56430}k_3 - \frac{9}{50}k_4 + \frac{2}{55}k_5\right), (11.8)$$

where $\hat{y}_n$ is the value of the previous, successful step. As stated we can estimate the local error in the step by finding the difference between Equations (11.8) and (11.7). Performing this operation and after some rearrangement we arrive at the following expression for the error in the local step

$$\sigma \equiv \hat{y}_{n+1} - y_{n+1} = h\left(\frac{1}{360}k_0 - \frac{128}{4275}k_2 - \frac{2197}{75240}k_3 + \frac{1}{50}k_4 + \frac{2}{55}k_5\right).(11.9)$$

Note that with this expression for the estimate of the local error we don't have to calculate the expression for y_{n+1}, Equation (11.7), in order to compute σ. Of course, $\hat{y}_{n+1}$ still needs computing.

Seeing as we know how the error behaves with step size (*fourth order Runge-Kutta*) we should be able to use this information to estimate the next step size from the step we have just computed. To do this we note that we can write

$$\frac{|h|\varepsilon}{|\sigma|} = \left(\frac{h'}{h}\right)^4 \tag{11.10}$$

where ε is the global error tolerance we want in our solution thus $|h|\varepsilon$ is our desired local error tolerance*, h' is (an estimate of) the "ideal" step that will produce the desired local error tolerance, and h is the step size for which we have just calculated σ.

Rearranging Equation (11.10) for the "ideal" step size we obtain

$$h' = (\alpha h)\sqrt[4]{\frac{|h|\varepsilon}{|\sigma|}}, \tag{11.11}$$

where we have included a parameter α as a factor we can adjust to provide a more conservative estimate of the next step, and thus takes values in the range [0,1]. Remember that Equation (11.11) is an *estimate* of the "ideal" step found from considering the error behaviour as the *order* of the fourth power of the step size *not h^4* itself. Erring on the side of caution we assume that it produces a step size that is (slightly) larger than the "ideal" thus warranting the inclusion of a conservative factor. Typically $\alpha \approx 0.9$ but can be adjusted if we find the error tolerance in the results to be unsatisfactory.

We should also be conservative in our approach to adapting the step size and as before we introduce maximum and minimum step sizes in order to take account of any singularities, discontinuities, or asymptotes, i.e. where the differential equation may change in nature. In addition to these limits we should also be conservative by how much the step size changes from one step to the next. If we find that the algorithm wants to increase the step size by more than a factor of ten, say, then we should be cautious and limit the increase

---

* The absolute value takes into account backwards integration

to a factor of ten (or less). Similarly, if the step size is suddenly decreased by a large factor we should also be cautious here and set an appropriate limit. These four limits (maximum and minimum step sizes, and the factors of maximum increase and decrease) should be experimented with for different, differential equations as, if you'll excuse the pun, one-size does not fit all, and it is likely we are unable to predict the nature of the (numerical) solution to the differential equation before-hand. One general strategy that might be applied is to make the maximum and minimum step sizes some fraction of the integration interval.

The file *RKF45N_mod.f90* contains the module code to run the algorithm described above for both first ordered ODEs (subroutine RKF451) with only one dependent function and second ordered ODEs (subroutine RKF452N) with any number of dependent functions; you will have to write a program that uses this module and calls the subroutine(s). Read through the code and make sure you understand each variable name, each line of the subroutines and the task they perform. Here we have defined each of the coefficients in Equations (11.1) through (11.9) as parameters to ensure that they are not changed by the program, and to make the code more readable. Also we have left them as fractions to ensure the best possible precision. Additionally, all the conservative precautions we have taken are given as adjustable parameters in the top few lines of both subroutines so they can be easily changed if necessary.

To check that the algorithms behave as expected and produce tolerable errors in their numerical solutions write a program to perform simple harmonic motion. Does the algorithm adapt the step size as expected? Does the algorithm remain numerically stable, and if so for how long? Indefinitely perhaps?

Before moving on to apply our Runge-Kutta-Fehlberg algorithm to some physics problems I first what to discuss an alternative method of viewing the dynamics of a system...

11.2 Phase Space

Normally, and quite intuitively, we plot the state of a dynamical system, e.g. the displacement of a pendulum, as a function of time. This will quite naturally tell us things like the amplitude of the oscillations and their frequency. However, we can create a plot that has no explicit dependence on time, one in which we plot the position of the pendulum, say, against its velocity (or more generally its momentum). This plot is known as the phase space of the dynamical system. As time advances, a point in phase space representing the current *phase state* of the dynamical system will shift, tracing out a *phase trajectory*. When we plot several phase trajectories for different initial conditions, say, or for different parameters we call that a *phase portrait*.

To illustrate, consider the motion of an undamped, mass-on-a-spring undergoing simple harmonic motion. We all know that the displacement of the mass is described by a sinusoidal function and that this function lags that describing the velocity of the mass by one quarter of a cycle. In other words, if $x = \sin(t)$ then $\dot{x} = \cos(t)$. Plotting the velocity $\dot{x}$ against the displacement x we would obtain a circle in phase space. This circle tells us that the motion must be oscillatory and also that the energy of the system is conserved. What would happen to the phase portrait if $x = \sin(\omega t)$ for $\omega \neq 1$? What affect does changing the amplitude of the oscillations have on the phase portrait? How might the phase space portrait look if we were considering damped oscillations, e.g. energy lost through air resistance? You should by now be thinking of how you can show your answer to those questions to be correct using the (computational) tools at your disposal.

For reasons that should now be obvious the origin of the phase portrait of the *damped* simple harmonic oscillator is called an *attractor*. The phase trajectory *spirals into* the origin as the oscillator *looses* energy. If we were to follow that trajectory in reversed time then we would see that the phase state of the system spirals out from the origin. In other words the oscillator is gaining energy and thus, in this case, we must have driving force.

We have already seen in Chapter 6 that when both driving and damping forces are present the oscillatory system goes through a transient phase before settling into a steady state. The nature of the transient phase is dependent upon the initial conditions of the system, whereas the steady state is not. How does this look on a phase space portrait, and what does *limit cycle* mean? How would resonance be detected using phase space?

11.3 Van der Pol Oscillator

Balthasar Van der Pol was a Dutch physicist and electrical engineer who experimented with some novel electronic circuits containing triodes (vacuum tubes) in the 1920s. One of those circuits is now know as the Van der Pol oscillator that is described by the non-linear differential equation

$$\ddot{y} = -y - \mu(y^2 - 1)\dot{y}, \tag{11.12}$$

where y is some position coordinate as a function of time, and μ is a parameter indicating the strength of the non-linear damping term. When $\mu = 0$ we have simple harmonic motion. Equation (11.12) describes self-sustaining oscillations in which energy is fed into small oscillations and removed from large oscillations; to see this consider the damping term when $y > 1$ and when $y < 1$.

11.3.1 Van der Pol in Phase Space

The Van der Pol oscillator equation is difficult to solve analytically due to the non-linear damping term but can be tackled using perturbation theory. However, this only works when μ is small i.e. $\mu \ll 1$, which is hardly interesting at all. This is where our numerical integrator steps in to provide the solution.

Figure 11.1 shows the phase space portrait of the Van der Pol oscillator for various values of the parameter μ. To obtain these plots we set the initial values of the displacement and the velocity to 2 and 0 respectively, and the tolerance was set to $\varepsilon = 10^{-5}$. Here we show that for $\mu = 0$ we recover simple harmonic motion. Note that the initial values were chosen so that they lay on the limit cycle for the oscillations.

Here we can clearly see the affect of the non-linear damping term on the phase trajectory of the oscillator. You may find it illustrative to also plot the regular displacement-time graphs to match up corresponding points from the phase portrait.

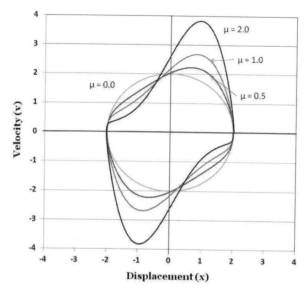

Figure 11.1: Phase portrait for the Van der Pol oscillator for various values of μ.

11.3.2 Van der Pol FFT

Normally we use Fourier transforms to gain (extra) insight into experiment data that is some function of time, say, by converting it into a function expressed in terms of frequency. We can also apply this analysis to numerical data, such as that which has be synthesised through computation.

We have just seen that the phase space diagram of the Van der Pol Oscillator is a alternative, and useful way to study the behaviour of the oscillator. The Fourier transform of the data will augment our understanding of the that behaviour.

If you recall, the Fast Fourier Transform (FFT) subroutine requires (time) data that is set a constant increments but our Runge-Kutta-Fehlberg solver is an adaptive step algorithm. Rather than try to interpolate the data from the solver we shall instead modify the routine to store values at constant time intervals. We can then pass this data directly to the FFT subroutine without the need to change it in any way.

The file *rkf45_FFT.f90* contains the module code to implement the modifications required. In essence, we check to see if we have reached our goal of the time increment we require. If we have, we increment our data counter, write the data to an array, and move the goal to the next desired data point. If we haven't reached our goal yet we continue the integration without storing any data. However, before moving on we include a check to see if the next step will take us past our current goal, and adjust the step accordingly, i.e. reduce the step such that it will hit the goal. In this way we should maintain an error that is less than the desired tolerance. As these oscillations tend to have an initial transient phase before settling into a steady state we should, in general, set a non-zero initial goal to ensure we start taking data from the steady region. We run the integration until we have filled our data array, which we can then pass to the FFT subroutine for analysis.

When sampling the data in this way we must remember that the FFT subroutine works best when the time increments are commensurate with the time period of oscillations. Put another way, we should ensure that we are not aliasing our data, and we fit an (exact) integer multiple of oscillations in our data array. We recall that our time increment (or sample rate) is given by

$$\Delta t = \frac{kT}{N}, \qquad (11.13)$$

where k is some integer, T is the time period of the oscillations, and N is the number of data points we will use in the FFT subroutine.

By far easiest way to obtain T is to use the displacement-time graph to estimate a value; take the time period for several oscillations and divide through by this number. For $\mu = 2$ I found the time period of oscillations to be around 7.631 seconds. As we are taking 512 data points (this number is *not* arbitrary; a power of two is required for the FFT subroutine) we should sample with time increment of, say, $\Delta t = 0.23846875$ seconds to include 16 full oscillations in our data array. This choice in the number of oscillations is again not arbitrary. By choosing a power of two we get an exact integer number of points per oscillation (give or take some small error in the calculation of the time period), in this case $2^9/2^4 = 2^5 = 32$.

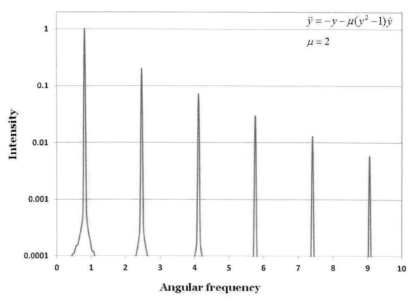

Figure 11.2: Fourier spectrum of the Van der Pol Oscillator, with damping parameter $\mu = 2$

You will know if your calculation of the time period is accurate by the quality of the Fourier spectrum. Figure 11.2 shows the Fourier spectrum for the Van der Pol Oscillator with $\mu = 2$, simulated with our RKF algorithm with a tolerance of 10^{-5}. Note that the intensity axis is logarithmic such that the decrease in intensity from one peak to the next is not linear but exponential. Here we can see we have picked a good time increment as the peaks are sharp, almost delta functions, and the "background" spectrum is small (less than 10^{-4}). The broadening of the peaks is most likely due to bin leakage as the time period we have calculated is close to but not quite exact. It may also be caused in part by the numerical errors in simulating the oscillations. Of course, this can easily be investigated by making changes to the tolerance in the RKF algorithm, and small changes to the time period.

11.4 The "Simple" Pendulum

You all probably remember the simple pendulum experiment from your physics classes at school. It provides a practical introduction to how to deal with experimental errors (time several oscillations to obtain a more accurate measure of the time period) and how to use approximation to simplify the maths. We can derive the differential equation for the pendulum from the geometry of the system such that

$$\ddot{\theta} = -\frac{g}{l}\sin\theta, \tag{11.14}$$

where θ is the angular position of the pendulum, g is the strength of gravity at the Earth's surface, and l is the length of the pendulum. Here we assume that the oscillations are free, in that there is no driving force (other than gravity), and there is no damping due to frictional or resistive forces. For the rest of this section let's also assume the pendulum is rigid. As Equation (11.14) is rather difficult to solve analytically (if not impossible?) the usual trick is to assume the small angle approximation that is $\sin\theta \approx \theta$ for $\theta \ll 1$ (in radians), and we obtain the simple harmonic oscillator equation. However, our numerical solver has no issues tackling this equation head on.

11.4.1 Finite Amplitude

With our new description of phase space we should be able to clearly visualise the behaviour of the pendulum, specifically seeing at what angles the small angle approximation holds. For small angles we should see a circular phase trajectory that will morph into something different as we increase the amplitude of the oscillations.

We have two approaches to consider in how to vary the amplitudes. We can either mimic what we would do given a physical pendulum. That is, we monitor the trajectory by directly varying the initial angle of release, and setting our initial velocity to zero. Or we consider the total energy of the pendulum, that is its potential energy plus its kinetic energy, and work out the (angular) velocity of the pendulum as a function of the total energy when the angle is zero, i.e. at the bottom of the swing, and use that as our initial conditions. We then vary the total energy (which will vary the amplitude) to see what affect this has on the phase trajectory. This second method is more practical in terms of phase space as the area encompassed by the phase trajectory is proportional to the energy in the system.

We recognise for any (mechanical) system the total energy is given by

$$E = T + V, \tag{11.15}$$

where

$$T = \frac{1}{2} m l^2 \dot{\theta}^2 \tag{11.16}$$

is the kinetic energy and

$$V = mgl(1 - \cos\theta) \tag{11.17}$$

is the (gravitational) potential energy. When the pendulum is at the bottom of its swing we have $\theta_0 = 0$ and

$$E = T = \frac{1}{2} m l^2 \dot{\theta}_0^2 \tag{11.18}$$

as this is where we have defined our zero potential energy. Rearranging Equation (11.18) for angular velocity yields

$$\dot{\theta} = \sqrt{\frac{2E}{ml^2}}. \tag{11.19}$$

Using units such that $g = l = 1$, and $m = 0.5$ we obtain the phase portrait of the simple pendulum as show in Figure 11.3. It is of note that using these units we set out unit of time as $\sqrt{l/g}$. Here we have computed the phase trajectories for total energies of 0.25 to 1.5 in steps of 0.25. Here the units of energy are dictated by those we chose so as to make the other parameters simple. As expected with lower energy (smaller amplitude) the pendulum behaves approximately like a simple harmonic oscillator. As we increase the energy (larger amplitude) that approximation no longer holds with the phase trajectory, elongating along the θ axis. The phase trajectory when $E = 1$ is called a *separatrix* as we can see from the trajectories that this defines a fundamental change in behaviour of the pendulum. What is the physical significance of $E = 1$, in other words what causes this change? Hint: The angular velocity of the phase trajectory reaches zero at $\theta = \pi$.

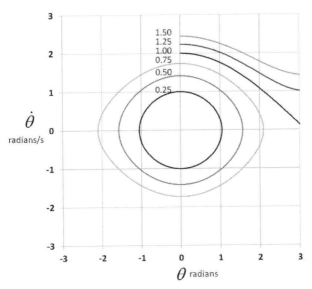

Figure 11.3: Phase portrait of the simple pendulum for various energies (amplitudes).

11.4.2 Utter Chaos?

Now that we have the phase description of the simple pendulum under our belts let's consider a more realistic system. As with the mass-on-a-spring system we introduce both a driving force, F_D, and a resistive, drag force, F_R, into our equations. The differential equation governing the motion of the pendulum is then given by

$$\ddot{\theta} = -\frac{g}{l}\sin\theta + \frac{F_D}{ml} + \frac{F_R}{ml} \tag{11.20}$$

where we have introduced the mass of the pendulum m into our equation. Here we consider that the mass of the pendulum is located at the very end of its length. To keep things simple (relatively speaking) let's assume the driving force is described by a periodic function such that

$$F_D = f_0 \cos(\omega_0 t), \tag{11.21}$$

where f_0 is the strength of the force and ω_0 is its frequency, and the resistive force can be described by

$$F_R = -\rho v = -\rho l \dot{\theta}, \tag{11.22}$$

where ρ is the coefficient of the drag force, and v is the tangential velocity of the pendulum mass.

If we rewrite Equation (11.20) in a dimensionless form, where we again choose $g = l$ (not necessarily equal to one) such that the unit of time is $\sqrt{l/g}$ we obtain

$$\ddot{\theta} + q\dot{\theta} + \sin\theta = b\cos(\omega_0 t), \tag{11.23}$$

where $q = \rho/m$ and $b = f_0/ml$ are adjustable parameters, along with the driving frequency ω_0.

Depending on the relative values of p, b, and ω_0 the motion of the pendulum can either be periodic or chaotic as shown in Figure 11.4. The discontinuities arise because we map the angle back into the physically valid range. When writing a program to drive this simulation we have to remember that our angle θ can only exist in the range $[-\pi, \pi]$. That is, we should confine θ to this range by applying the transform $\theta \pm 2n\pi \rightarrow \theta$, where n is determined from the ratio of $\theta/2\pi$. The most straight forward way of doing this is to modify our RKF algorithm; after we have accepted the step and have assigned YHAT to Y0, we adjust the value Y0(1) (θ in this case) by adding the following line

Y0(1) = Y(1) - 2* PI*NINT(Y(1)/(2*PI)).

The intrinsic Fortran function 'NINT' rounds the argument to the nearest whole number with the policy that 0.5 goes to 1. You can always *comment out* this line after you've added it should you be investigating a problem that is *not* confined to the range $[-\pi, \pi]$.

I'll leave it as an exercise for the reader to investigate the relative values of the parameters required to bring about chaos. Note that whole books are dedicated to the study of chaos and chaotic motion, and it is still under much academic research. One thing to remember you could always analyse its Fourier spectrum...

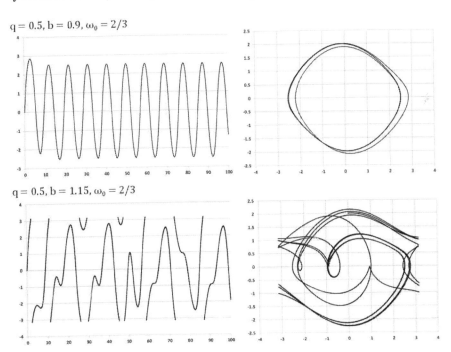

Figure 11.4: The angle as a function of time and the phase trajectories of the driven pendulum showing periodic motion (top) and chaotic motion (bottom) for the parameter values shown.

11.5 Halley's Comet

Halley's Comet is probably the best known short-period comet and is visible from Earth with the naked-eye every 75 to 76 years. Halley's returns to the inner Solar System have been observed and recorded by astronomers since at least 240 BC. Clear records of the comet's appearances were made by Chinese, Babylonian, and medieval European chroniclers, but were not recognized as reappearances of the same object until much later. In 1705 English astronomer Edmond Halley was the first to calculate the comet's periodicity, and was rewarded with having it named after him. Halley's Comet last appeared in the inner Solar System in early 1986.

Halley's Comet has a highly elliptical, planar orbit with large differences in its velocities at the aphelion (furthest distance) and the perihelion (closest distance) of its journey around the Sun. The equation governing the comet's trajectory (i.e. the force, F, acting on the comet) is Newton's Law of gravitation,

$$F = -\frac{GMm\hat{e}_r}{r^2} = -\frac{GMm\underline{r}}{r^3}, \qquad (11.24)$$

where G is the universal gravitational constant, M is the solar mass, m is the mass of the comet, $\hat{e}_r = \underline{r}/r$ is a unit vector that points from the centre of the Sun to the centre of the comet, and r is the distance between the centre of the Sun and the centre of the comet. Here we assume that the influence of other bodies in the solar system are insignificant compared to the gravitational pull of the Sun.

We usually give the units of Equation (11.24) in SI form. That is, distance is measured in metres, time in seconds, and mass in kilograms. This makes the universal gravitational constant $G = 6.67384 \times 10^{-11} \text{m}^3\text{kg}^{-1}\text{s}^{-2}$, the solar mass $M = 1.9891 \times 10^{30}$ kg, and the aphelion distance is 5.28×10^{12} m. Using the vector form of Equation (11.24) we have to calculate the distance cubed. This is going to lead to precision problems if we use SI units and a change of units to more computer friendly ones is necessary. The first thing to note is that both G and M are constants thus we can write $G_S = GM$ as the universal gravitational constant *per solar mass*. We now need to choose the units for length and time so that both the solar-comet distance r, and our gravitational constant per solar mass G_S have exponents that ideally reduce to zero, and certainly no more than one. Instead of arbitrarily choosing some units let's use some that are more natural. The astronomical unit, AU, defines the mean distance between the Earth and the Sun, which has a value of $1.49597871 \times 10^{11}$ m. This makes the aphelion distance equal to 35.1 AU so this appears to be a good choice; remember this is the greatest distance from the Sun.

The sidereal year is defined as the orbital period of the Earth around the Sun relative to the background of "fixed" stars. The name sidereal comes from the Latin 'sidus' meaning 'star'. It is used often in astronomy[*] and contains 365.256363 days equal to 3.1558150×10^7 s. So if we define our computer friendly units of length, mass, and time as the astronomical unit (AU), solar mass (M), and (sidereal) year (yr) respectively then $G_S = 39.489$ AU^3 M^{-1}yr^{-2}. (To obtain this value you multiply G in SI units by M, then multiply by the square of the number of seconds per year, finally diving by the cube of the number of metres per AU).

To solve Equation (11.24), which is a second ordered ODE, and thus determine the trajectory of Halley's Comet we need to know two pieces of (initial) information. That is, we need to know the comet's position and (instantaneous) velocity at a particular time, which we take to be our origin in time. As we already know the aphelion distance we can use that position to start the integration; the aphelion velocity is 912 ms^{-1}. After making the appropriate changes to the units and writing a program that uses the RKF algorithm for a seconded ordered ODE with *two* dependent functions (x and y coordinate of the comet) we plot the comet's trajectory in Figure 11.5. Figure 11.5(a) shows the trajectory for as single orbit where each data point is taken one year apart; note the difference in the scales of the x and y axes. Figure 11.5(b) shows the *distance* from the sun plotted as a function of time for several orbits; again the data points are one year apart.

---

[*] And as a way to confuse students.

Note that if you tried to solve this differential equation using a constant step length algorithm you will find that unless the step length is very small, the solutions are very unstable orbits with the comet shooting off somewhere as it approaches and goes past the Sun – this clearly doesn't happen.

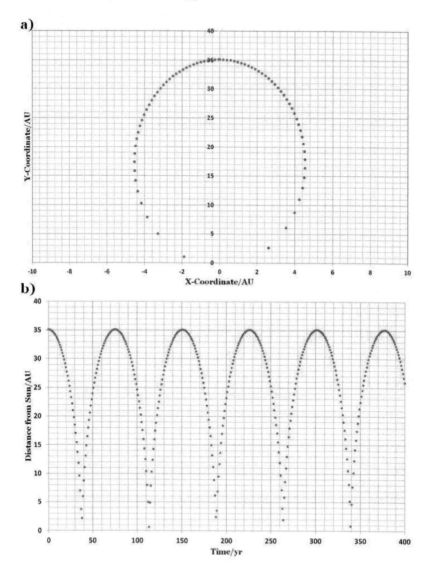

Figure 11.5: Trajectory of Halley's comet (a) looking down on the orbital plane, where the Sun is located at the origin; (b) comet's distance from the Sun as a function of time. In both the points shown are spaced one year apart.

11.6 To Infinity and Beyond...

Humans have long wondered what's out there amongst the stars. Exploration, it seems, is in our nature. Our first goal is to get to our nearest neighbour in the solar system; the Moon. The first obstacle to overcome is how we get off the planet in the first place. That I leave as a task for the reader (see Exercise 5). Assuming we've achieved a stable orbit about our planet, our next obstacle is to navigate to the moon. Unlike the movies, we do not have the luxury of an endless supply of fuel and are reliant on short burst thrusters only, meaning that the motion of our spaceship is (mostly) dictated by Newton's Law of gravitation. The speeds we consider are nowhere near relativistic, neither are the gravitational forces, such that Newton's Laws are an adequate description of the physics. We pick our frame of reference as the Earth-Moon system; this frame of reference is in orbit about the Sun and as such we can consider the relative motion of the Earth, Moon, and spaceship independently from their motions about the Sun (and the Sun's motion about the galaxy, the galaxy's motion about the local cluster, and ...). With these descriptions in place let's go to the Moon.

As a start let's just consider the Earth-Moon system. Normally we state this as the Moon orbiting the Earth but in fact they orbit each other about some, common centre-of-mass (COM). Note that this true of any two-body system orbiting one another. It makes sense, therefore, to fix our origin at this centre-of-mass. In general, orbital trajectories are elliptical with one of the foci located at the origin of the system. However, the eccentricity of the Earth-Moon orbit is sufficiently small that their trajectories can be assumed to be circular. In addition to this their orbits are planar, that is they can be sufficiently described by two spatial coordinates. If d is the centre-to-centre distance between the Earth and the Moon then the distance of the centre of the Earth to the COM is

$$r_E = \frac{m_M}{m_M + m_E} d \qquad (11.25)$$

and the distance of the centre of the Moon to the COM is

$$r_M = \frac{m_E}{m_M + m_E} d, \tag{11.26}$$

where m_E is the mass of the Earth and m_M is the mass of the Moon. If you're wondering how we arrive at these equations then the trick is to consider turning moments.

The Earth-Moon system rotates about its common centre of mass with a sidereal orbital period T. If the Moon lies on the positive x-axis at $t = 0$, then

$$\phi_M = \omega t \tag{11.27}$$

and

$$\phi_E = \omega t + \pi \tag{11.28}$$

where ϕ_M is the angular location of the Moon, ϕ_E is the angular location of the Earth at time t, and ω is the angular frequency of the orbit; the angle is measured from the x-axis. To remain in circular motion a body must be constantly accelerated toward the centre of motion, with an acceleration of magnitude $\omega^2 r$, where r is the length of the radius of the motion. For the Earth that acceleration is provided for by the gravitational force between the Earth and the Moon such that

$$\frac{G m_M}{d^2} = \omega^2 r_E. \tag{11.29}$$

After substitution of Equation (11.25) and some manipulation we arrive at Kepler's third law for planetary motion

$$\omega^2 = \frac{4\pi^2}{T^2} = \frac{G(m_M + m_E)}{d^3}. \tag{11.30}$$

We would arrive at the same relationship if we had first considered the acceleration of the Moon. Rigorously speaking this is not really proof of Kepler's third law as we have assumed circular orbits and more generally we should consider elliptical orbits. However, Equation (11.30) does hold for elliptical orbits but in this case d

would be the semi-major axis of the ellipse rather than the centre-to-centre distance.

Equations (11.25) through (11.30) now form a practical description of the relative motion of the Earth and Moon about each other. As stated our spaceship is currently in a stable orbit about Earth. Let's assume that this orbit is 500 km *above the surface* of the Earth. When the spaceship reaches some particular angular location θ in its orbit, it fires its thrusters, and accelerates to some speed v in a direction tangent to the orbit at θ. Here we'll assume that this thrust acceleration is instantaneous in comparison to the total journey time; you will see that this is a reasonable assumption once we perform the computations. Our spaceship is now in motion towards the Moon – we hope! But as the spaceship travels, so do the Earth and the Moon move about their COM, and our spacecraft in influenced by their gravitational fields such that

$$\underline{F} = m\underline{a} = -Gm\left[\frac{m_E}{(r-r_E)^3}(\underline{r}-\underline{r}_E) + \frac{m_M}{(r-r_M)^3}(\underline{r}-\underline{r}_M)\right], \quad (11.31)$$

where m is the mass of the spaceship, which neatly cancels from our equations, and $\underline{r}$ is the position vector of the spacecraft. Writing these in component form for the x and y directions we have

$$\ddot{x} = -G\left[m_E\frac{x-x_E}{d_E^3} + m_M\frac{x-x_M}{d_m^3}\right], \quad (11.32)$$

and

$$\ddot{y} = -G\left[m_E\frac{y-y_E}{d_E^3} + m_M\frac{y-y_M}{d_m^3}\right], \quad (11.33)$$

where the distance of the spaceship from the centre of the Earth is given by

$$d_E^2 = (x-x_E)^2 + (y-y_E)^2, \quad (11.34)$$

and the distance of the spaceship from the centre of the Moon is given by

$$d_M^2 = (x - x_M)^2 + (y - y_M)^2. \tag{11.35}$$

The x and y components of the Earth and the Moon distances from the COM are given by

$$x_E = r_E \cos(\phi_E), \quad y_E = r_E \sin(\phi_E) \tag{11.36}$$

and

$$x_M = r_M \cos(\phi_M), \quad y_M = r_M \sin(\phi_M). \tag{11.37}$$

As a task to the reader: find out the required physical constants you will need in order to compute the spacecraft's trajectory for different θ and v. This list consists of the mean centre-to-centre distance of the Earth to the Moon (d); the mass of the Earth (m_E); the mass of the Moon (m_M); and the *sidereal* Earth-Moon orbital period (T). We already know G from previous sections in this chapter. Is there anything else we should know? (Hint: the position vector of the spaceship is computed as the distance from the *centre* of the Earth and the Moon, and these bodies are certainly not point masses!)

Once we have discovered the necessary physical constants we're ready to compute. Or are we? Remember that we need computer friendly units such that we're not dealing with numbers that have large variation in their exponents. The strategy to employ here is as before with Halley's comet; to use the physical constants you have found as the units of measure. For instance, we would use the Earth-Moon distance as the unit of length, the sidereal orbital period as the unit of time, and the mass of the Earth as the unit of mass.

After performing the necessary changes to units you should notice that the value for the gravitational constant is very similar to the one found for Halley's Comet. This is not a coincidence and do you know why?

Once you have a program written to find the trajectory of the space ship you should check there is are no bugs in your code, such as incorrect entry of a physical constant, or a mistake in the change of units, and so on. To do this set the mass of the moon to zero and check that you get a stable, circular orbit of the spaceship about the Earth when you set the necessary velocity; you will have to use a variant of Equation (11.29) to work out the velocity required. If you get a circular orbit then we're ready to attempt to make that trip to the moon.

To monitor the progress of the spacecraft we should store the distance to the centre of the moon for evaluation, and perhaps terminating the program once we are at or within the radius of the Moon – obviously this means we've likely collided with the Moon, but landing on the moon is another problem to solve. I leave it as an exercise for you to do to find values for θ and v that will get the spaceship to the moon.

For more animated applications of ODE solvers to BIG physics check out 'Universe Sandbox'.

11.7 To the Infinitesimal and Below...

Back in Chapter 4 we discussed how to obtain the solutions of the Schrodinger Equation as applied to the infinite square well and the finite square well. In the infinite square well case we found the solutions analytically, whereas for the finite square well we had to rely on root finding to provide the energy eigenvalues. With our Runge-Kutta-Fehlberg ODE solver we should be able to tackle any arbitrarily defined electrical potential function with ease. Or so you would think (or not depending on how seasoned you are as a physicist).

As a starting point, let's see if we can emulate the results we obtained for the finite square well using root finding with our adaptive ODE solver. Rather than starting entirely from scratch let's use the energies found from the root search applied to the functions

$$f(E) = \beta \cos(\alpha a) - \alpha \sin(\alpha a) = 0, \tag{11.38}$$

for the even parity states and

$$f(E) = \alpha \cos(\alpha a) + \beta \sin(\alpha a) = 0, \tag{11.39}$$

for the odd parity states, and plug those into our differential equation. Here we are assuming we don't know the form of the solution of the wavefunction, instead we are relying on our integrator to provide us with the answer. As such we need to provide our integrator with a starting point. We could use the middle of the well were we know from experience that even parity states have $\psi(x) \neq 0$ and $\psi'(x) = 0$, and odd parity states have those relations reversed. We would then integrate from the middle of the well to the left, and then integrate from the middle of the well to the right to provide us with the full solution. However, this relies on the knowledge of the behaviour of the wavefunctions in the well, which in general we don't know, and in fact is why we are using the solver in the first place! We need a more general starting location.

We know that for any (arbitrary) potential the wavefunction vanishes to zero as we go deeper into a classically forbidden zone. Let's choose a starting location, x_0 that is deep into the barrier, left of the well, and integrate to the symmetrical position on the right of the well. At the starting location we can set

$$\psi(x_0) = 0. \tag{11.40}$$

It is also true that the derivatives of the wavefunction vanish the deeper we penetrate into the barrier. However, if we set $\psi'(x_0) = 0$ we would obtain a solution that implied the wavefunction was zero everywhere; we would have no particle in our system. Therefore, we set $\psi'(x_0)$ to some small, positive value; positive because we know that the probability of finding the particle in the barrier increases as we approach the boundary with the well. So how do we choose the magnitude of the starting differential? The answer is that the size really doesn't matter* from a qualitative point of view. All the size of the differential at the starting point does is *scale* the numerical solution of the wavefunction. If we chose $\psi'(x_0) = \delta$ and performed the integration, then changed the value of $\psi'(x_0)$ to 5δ, say, then our wavefunction from this integration would simply be five times that of the previous integration. The physically significant scale factor is the one that normalises the probability function such that

---

* that's what she said!

$$\int_{-\infty}^{\infty} \psi^*(x)\psi(x)dx = 1 \qquad (11.41)$$

but as we're only interested in the qualitative results for this discussion, that is another problem solve elsewhere.

On a practical note, try not to start the integration too deep into the barrier. You will find that if you do you then, even with values of $\psi'(x_0)$ on the order of the machine precision, our adaptive step integrator *will not* be able to cope with the change in nature of the differential equation as we cross the boundary between the barrier and the well. Not unless we relax the error tolerance significantly, and this would then give us doubts about the validity of our numerical results. For a 10 Å well, centred at the origin I found that a starting point of $x_0 = -8$ Å was about as deep as I could go without any difficulty (using double precision variables); here I set the derivative equal to the error tolerance I used for the root search and integrator, specifically 10^{-8}.

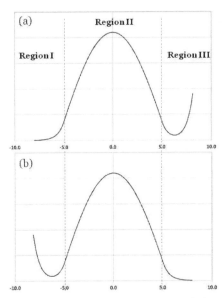

Figure 11.30: Results of the integrator: (a) integrating left to right; (b) integrating right to left.

Figure 11.6(a) shows the results of the integration as discussed above for the ground state function. We see that regions I and II have been computed correctly showing the same qualitative result as the root finding function. But what's going on in region III? We see that the wavefunction initially behaves as expected as we enter the barrier but as we go deeper it blows up exponentially! With a logarithmic vertical axis the increase is linear. This anomaly can also be seen in the higher energy states. As we are using an adaptive step solver with a low degree of tolerance (10^{-8}) then we can rule out numerical error as the cause. Also as the wavefunction behaved as expected in the other two regions we can rule out programming error with some confidence (though a check might be prudent in general). To gain further insight into the cause of this unphysical behaviour of the wavefunction let's perform the same integration in the reverse direction. That is, start at $x_0 = 8$ Å and integrate backwards to $x = -8$ Å. Here $\psi'(x_0)$ will now be some small, *negative* value. The results of this reverse integration are shown in Figure 11.6(b). Here we see the same problem but now in region I not region III! The fact that the backward integration looks like a reflection in the vertical axis of the forward integration lends credibility to our supposition that the programming is correct. Clearly the direction of the integration affects the numerical solution. But why, then, is the affect asymmetrical?

To answer that question we look toward the *general* solution for Schrodinger's equation in the barrier regions, specifically in region I we have

$$\psi_I(x) = Ce^{\beta x} + De^{-\beta x}.$$
(11.42)

Thus, mathematically speaking, within the barrier the wavefunction consists of two exponential terms; one grows while the other decays. Recall that we set D to zero using a physical argument based on observation. Thus we assumed the wave function had the form $Ce^{\beta x}$. However, mathematical equations tend to be oblivious to our physical reasoning. What does this mean for our integration? Let's consider the first integration. Here we progress the solution forward from a *negative value* of x such that in region I the *magnitude* of x decreases. This means that our desired wavefunction term $Ce^{\beta x}$ is the *growth* term, whereas the *unwanted* term $De^{-\beta x}$ is the *decay* term; take your time to verify this. Hence, we are integrating in the direction where the unwanted term decays.

In region III of the forward integration we start at a *positive* value of x, namely the well boarder, and progress from there such that the *magnitude* of x *increases*. From symmetry arguments the general solution to Schrodinger's equation for region III is the same as for region I (with a coefficient sign reversal for odd parity wavefunctions). In this case the desired solution is the $De^{-\beta x}$ term, and the unwanted term is the $Ce^{\beta x}$. In other words, our desired solution decays while the unwanted term grows. Thus we have identified the cause of our problem. We can apply similar arguments to the backward integration and find that while the unwanted term decays in region III, it grows in region I. The reason why the unwanted terms exists in the first place it because of the slight imprecision in the calculation of the energy eigenvalue. Even though we've calculated it using a root searching algorithm to a precision of at least the order of 10^{-8} it is *not* an exact value. Thus the coefficient of the unwanted term is not exactly zero, but some very small number. However, the exponential terms grows rapidly with x; exponentially in fact! Eventually, there will come a point where this small coefficient multiplied by the exponential growth factor will become the dominant term and cause our solution to blow-up where we would expect it to decay.

The remedy then is to always integrate from a classically forbidden region towards a classically allowed region. In this way any unwanted solution will decay, while the desired solution grows. For any symmetrical potential this is particularly easy as we can simply integrate from the left of the well to the middle of the well for various energies. We find the eigenvalue by finding the energy E that satisfies

$$\psi'(x=0,E)=0, \qquad (11.43)$$

for even parity wavefunctions, and

$$\psi(x=0,E)=0, \qquad (11.44)$$

for odd parity wavefunctions. The rest of the wavefunction will just be the mirror image of that calculated reflected in the vertical axis at the middle of the well; see Figure 11.6 to convince yourself of this.

A slightly better way of searching for the eigenvalue, in that it removes the ambiguity in the choice of $\psi'(x_0)$, is to search for the energy that satisfies the logarithmic derivative being zero, i.e.

$$\left.\frac{\psi'(x,E)}{\psi(x,E)}\right|_{x=0} = 0, \tag{11.45}$$

for even parity wavefunctions, and the inverse of this for odd parity wave functions.

The file *finiteWell2.f90* contains (less than optimal) code that can find at least one even parity, energy eigenvalue of the user defined (symmetric) potential $V(x)$; this is currently set to the finite square potential. The program structure is essentially the hybrid Bisection-Secant root searching algorithm with the RKF45N subroutine playing the role of the "function evaluation" and we are trying to find the energy that will satisfy Equation (11.45). You should think about restructuring this code so that the main program only calls the hybrid root search subroutine, and the RK45N subroutine is called from within the function we pass to the root search. In addition to this you will have to design a way to deal with the different parities, and design a way to bracket the roots for the logarithmic derivative search in the first place.

Be aware that although useful for instruction, symmetric potentials rarely arise in real quantum mechanical systems, and as such will not contain pure even and odd parity wavefunctions. However the general strategy of solution still applies; choose a matching point in the classically allowed region (i.e. the "well") ; integrate up to this point from opposite sides in the classically forbidden regions; and compare the logarithmic derivative at the matching point. Mathematically, we find the energy eigenvalues that satisfy

$$\left.\frac{\psi'_L(x,E)}{\psi_L(x,E)}\right|_{x=x_m} = \left.\frac{\psi'_R(x,E)}{\psi_R(x,E)}\right|_{x=x_m}, \tag{11.46}$$

where ψ_L is the numerical solution for the wavefunction integrated from the left to the matching point x_m, and ψ_R is the numerical solution for the wavefunction integrated from the right to x_m. As the notion of pure even and odd states doesn't apply Equation (11.46) holds for *any* energy eigenvalue. Note that it may happen that at the matching point we choose the wavefunction tends to zero and we end up with a singularity in the calculation of the logarithmic derivative. How might this be avoided in general?

So there you have it, we have a robust ODE solver that can tackle problems on the scale of the universe to the scale of the quantum to a user defined precision. If that's not awesome then I don't know what is.

Exercises

1. Investigate the Van der Pol oscillator further through variation of the damping parameter μ and the initial conditions. Can the Van der Pol oscillator ever become chaotic?

2. Establish a relationship between the driving frequency and the period of oscillations for a periodic (i.e. not chaotic) driven pendulum for set values of q and b. Is there a more general relationship as we vary q and b, but still within the non-chaotic region?

3. Duffing's oscillator is described by the differential equation

   $$\ddot{x} + \alpha\dot{x} + \beta x^3 + \gamma x = \delta \cos(\omega t)$$

   where α through δ are constants, and ω is the frequency of the driving force. Investigate the motion of the oscillator for different relative values of these constants.

4. Find the initial values for θ and v in order for our spaceship to loop the Moon and return to Earth.

5. Model the motion of a rocket that is launched from the surface of the Earth and establishes a stable orbit at 500km above the Earth's surface. To produce the thrust the rocket burns fuel and propels the gases out of its rear end, such that the mass of the rocket changes with time. Additionally, the density of the atmosphere is a function of height above the Earth's surface and this should be taken into account.

6. Write a program to simulate a journey to our next nearest neighbour in the solar system, Mars. How precise should we make our calculations?

7. In realistic, solid state quantum well devices the potential walls of the well are better modelled by a graduated slope rather that an abrupt "cliff edge". Investigate the effect of the steepness of the sloped walls on the bound energy states of the well.

8. Investigate the bound states of the potential centred on the origin, defined by

$$V(x) = \begin{cases} V_1, & |x| > b \\ V_2, & a < |x| \leq b \\ 0, & |x| \leq a \end{cases}$$

where $V_2 > V_1 > 0$ and $|b| > |a|$. Comment on the states with energy eigenvalues greater than V_1 but less than V_2. (Tip: It would be extremely useful to sketch this potential before trying to solve for it computationally).

9. A symmetrical anharmonic potential in one dimension can be written as fourth ordered polynomial such that

$$V(x) = \alpha x^4 + \beta x^2.$$

Find the first four energy eigenvalues for $\alpha = 0.5$ and $\beta = 1.0$ to at least 6 significant figures of accuracy. Study the effect of different values of α and β on these energy eigenvalues. You should plot $V(x)$ with the wavefunctions computed, offset by the corresponding energy eigenvalue. Investigate the effect on the wavefunctions as we add odd powers of x to the potential. Note that when $\alpha = 0$ we have an *harmonic* oscillator.

12 High Performance Computing

"The important thing is not to stop questioning."

- Albert Einstein

In the other chapters of this book we have only looked at getting algorithms to work as computer code. In this chapter we look at getting algorithms to work quickly or efficiently – these are not necessarily the same thing. To that end this chapter explores two methods to achieve high performance computing. Firstly, blocking that simply makes efficient use of the computer's memory architecture and, secondly, parallelism that attempts to utilise the total potential computing power of multiple core processors.

This chapter will discuss some of the fundamental ideas of memory structure and memory access but is by no means exhaustive or comprehensive. As with all things in computing there are levels of abstraction, the more levels you peel away the more technical (and usually complex) the ideas get.

To compile the code in this chapter you will need to include the following flag to tell the complier you're using OpenMP (OMP) directives (these will be explained in due course):

gfc –fopenmp <program_name>.f90 –o <executable_name>

12.1 Indexing and Blocking

In this section we will discuss in more detail the underlying structure of your computer and how we as programmers can make best use of that structure. Note that the majority of what we will discuss can be handled automatically by most modern compliers. However, it is always prudent to be aware of how a computer is put-together and how it operates in order to ensure the best possible performance the hardware can manage.

12.1.1 Computer Memory

As discussed in the introductory chapter, each level of computer memory can be thought of as a huge filing cabinet, each draw representing a memory address in which we can store one word (remember a word is 4 bytes, or 32 bits long). The memory can only be accessed one draw or address at a time and the current address is referenced by the system's address pointer, generally referred to as the Program Counter. To change from one address to another the Program Counter can either step from address to the next, or can be instructed to jump. Think of it like changing the channel on your TV using the channel + and − buttons shifting to adjacent channels, or inputting the number directly and jumping to that channel. Computer's memory is commonly referred to as being contiguous; the address locations share a common border.

We also discussed in Chapter one that computer memory is split into a hierarchical system whereby the smallest memory (the CPU cache) is the fastest, and the largest memory (the HDD) is the slowest. RAM exists between CPU cache levels and the HDD. When performing an operation the CPU will ask for the variables that require work. If the variables are not already in cache a signal is sent to fetch them from RAM. If the variables are not in RAM then a signal is sent to fetch them from the HDD. The variables are then read and copied from the HDD into RAM, then read and copied in the CPU cache levels. Once in cache the CPU performs the required operation and writes the result and/or changes to the variables back to RAM, which in turn writes those changes back to the HDD. Each one of these stages requires at least one clock cycle to complete, and for accessing the HDD may require several hundred! Note that those variables will now persist in RAM and the CPU cache levels until they are flushed by the system. This persistence allows the CPU quicker access to those variables should they be required again in the near future.

Typically, programs need to work on arrays and will work on those arrays in a consecutive manner. For example, let's say we have two vectors, a, and b of length N that require addition. Code is written as a loop that iteratively steps through the arrays one element at a time, performing the addition. It would be terribly inefficient if the fetch instruction only brought up the two variables from storage that required immediate addition; the fetch instruction would have to be issued N times i.e. the HDD has to be accessed N times. Far more efficient would be to bring up a block of variables at a time temporally storing them to RAM then the cache, and if N is sufficiently large, filling the cache levels. FORTRAN stores arrays as unbroken blocks of memory with the array elements in consecutive order. Subsequent array variables can now be accessed quickly with fewer calls made to the HDD. The number of fetch instructions sent to access the HDD is now approximately the ratio of the array length to the block size. You may therefore think that the best block size would be the length of the array, however the block size is limited by the amount of data that can be passed via the memory buses (typically gold alloy wires) that connect the different memory components. Generally this block size, referred to as the cache line, will be some fraction of the size of the level 1 cache and will be some integer multiple of eight. For clarification, the cache line is measured in bytes rather than the actual number of variables contained in the line due to the differences in variable types. For example, if the cache line were thirty two bytes long this would be enough to store eight, four byte words (single precision) or four, eight byte words (double precision).

This idea of blocks of memory affecting the performance of your computer is one you may have come across before if you've ever defragmented your hard drive to make it run quicker. Programs store variables and data on the HDD in blocks of memory that are access when that program is run. During the lifetime of your computer those blocks become broken and jumbled up, in technical parlance fragmented. This has detrimental effect on your computer because the CPU has to issue more fetch commands in order to receive the correct pieces of memory. By defragmenting the HDD those blocks reform into unbroken pieces of memory, which helps improve the performance of your computer. It also has the secondary effect of freeing up some storage space.

Of course, while the CPU is busy performing the required operations on the variables now stored in level 1 cache the other components don't have to be idle. Other blocks of memory can be fetched up to fill level 2 cache, and once full, begin to fill RAM. As the blocks of memory are finished with they are written back down the memory hierarchy, flushed from the cache, and fresh ones moved into level 1 cache to be worked on. This process is known as pipelining and is continuously working away in the background during the operation of your computer, making it incredibly efficient at number crunching. Normally this efficiency is implicit; the computer just does its thing. However, a computer is only as clever as the program telling it what to do. Sometimes the requests we make of the computer are, to put it technically, bloody stupid...

12.1.2 Loopy Indexing

To illustrate my point have a look at the code *matrixAdd.f90*. Here we are adding two very large matrices and storing the result in a third matrix of the same size. The program does this twice, timing both operations, and outputting the times to the screen. Note that on the second time through I have swapped the row and column indices loops. Before compiling and running this code, would you expect the times to be the same?

For all of you that ignored that question you should have found a significant difference between the two times. At first glance this seems wrong; both methods are performing the same number of additions so why the difference? To answer this we have to know how Fortran stores multidimensional arrays in computer memory. As discussed above computer memory can considered as huge filing cabinet, but importantly, it is a one dimensional filing cabinet; it has only one, albeit very long, column of draws or addresses. How then do we fit a multidimensional array into a one dimensional column? We reshape the array so that in essence it becomes a very long one dimensional vector, for example an n-by-m matrix, becomes an $(n \times m)$-by-one vector. However, we have a choice about how the array elements get stored in memory; do we follow the row order or the column order? Fortran follows row order; the columns of the matrix are stored in contiguous blocks of memory, with the caveat that the whole matrix takes up one unbroken block of memory. C/C++ follows column order; the *rows* of the matrix are stored in contiguous blocks of memory, but with no condition that the whole matrix be stored in one unbroken block of memory*.

Combining the knowledge of how Fortran stores array variables with how the memory is fetched up the memory hierarchy gives us the reason why the performance of the two matrix addition methods differ. The first, quicker method has the row index, I, on the inner or nested loop. This means that while the row index increments by one each time round the loop, the column index, J, remains constant. Thus, we are scanning down the columns of the matrix, mimicking how the array variables are stored in memory, and taking advantage of the efficient pipelining. The second, slower method reverses those loops so that the column index is in the nested loop, and we now scan across the rows of the matrix. This change does two things. Firstly, the program counter now has to jump over the total number of rows of the matrix to find the next elements to add. Secondly, and most detrimentally, the variables that are fetched in the current cache line become redundant; they are flushed from cache memory before becoming useful as the matrix is larger than the cache size. In essence, we are forcing the CPU to call to cache only one element at a time for each matrix. This produces the loss in performance when swapping the nested loop index.

If you are building programs with high performance in mind then these kinds of memory structure considerations should be paramount. You should also consider how to keep total memory usage to a minimum. For instance, in the matrix addition example above do we really need to keep the source matrices in memory when all we want is the resultant addition?

---

* This is one of the subtle differences between Fortran and C++ of which you should be aware. Another is that array indexing in C++ starts at zero rather than one.

12.1.3 Blocking

Matrix multiplication is a little a more involved than matrix addition. Matrix multiplication is defined element-wise by:

$$c_{ij} = \sum_{k=1}^{m} a_{ik} b_{kj} \qquad (12.1)$$

for each i and j, where m represents the inner dimension of the matrix product, e.g. matrix A having dimensions n-by-m, matrix B having dimensions m-by-p, resulting in matrix C having dimensions n-by-p. Note that there are three indices namely i, j and k. Clearly, for large matrices (where the three matrices combined are larger than half the cache size) memory access will become a bottle neck for matrix multiplications. To see this effect for yourselves compile and run the program called *matixMul1.f90*. Here we generate matrices of increasing size, filled with double precision floats, and multiply them together using the naïve triple loop algorithm. Here the description naïve is not derogatory but refers to the fact that the algorithm can be rewritten to reduce the total number of floating point operations required to compute C. Note that the index loop order (i, j, k) is set for optimal memory performance for the naïve triple loop algorithm using Fortran. Go ahead a change them around if you feel the need to confirm this; I did.

The first step in writing a more memory efficient algorithm for matrix multiplication is to realise we can separate the matrices into sub-matrices of block rows or block columns. We then treat the whole matrix multiplication as performing multiplications using these strips. The strip width, i.e. the number of rows or columns contained within the strip, can be set so that the total amount of memory consumed when using the strips is equivalent to the cache size. However, there is a slight flaw in our strategy here. As the size of the matrix increases the width of our strips necessarily reduces to maintain a cache block size. Eventually a matrix of sufficient size will make the strip width equal to one and we have lost any performance improvement our strategy might have afforded us.

We therefore need to generate a sub-matrix whose size is independent of the size of the total matrix, and equivalent to the cache size. Taking the lead from the strips idea, whereby we divided the matrix along one of its dimensions, we now divide along the second dimension thus forming blocks. The easiest way of thinking about performing matrix multiplications with blocks is to treat the blocks as if they were elements. The row by column process still applies. Figure 12.1 illustrates this block multiplication process. Here the multiplication operations in the brackets can be done in any order, which is useful to know for parallel programming which we will discuss shortly. Using this process we can bring up two blocks into cache, one each from matrix A and B, perform a normal matrix multiplication, and store the result in the corresponding block of matrix C. The notation we'll use for the block, sub-matrices will be A_{nm} where n is the block row index and m is the block column index.

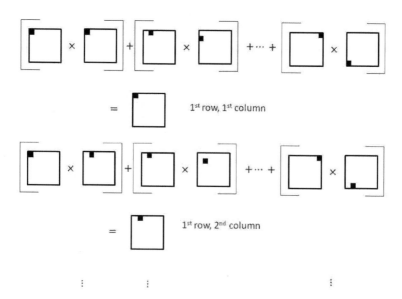

Figure 12.1: Block matrix multiplication. The bracketed terms are matrix multiplications in of themselves.

The idea with memory efficient programs is to ensure that the relevant variables, that is the variables we wish to work on, persist in the higher levels of memory (cache and RAM) until they have no further use and can be flushed. Looking more closely at the multiplication process we note that the first block of matrix A is only ever involved with the first block row of matrix B. In fact, only the first block *column* of A is involved with the first block *row* of matrix B. It is easy to extend this to the kth block column of A and the kth block row of B. This is actually a consequence of the inner product nature of matrix multiplication. Element wise we must have $c_{ij} = a_{ik}b_{kj}$ thus similarly for our sub-matrix blocks we must have $C_{ij} = A_{ik}B_{kj}$. Thus we might proceed by keeping A_{11} in cache while we iterate through the first block row of B. Then move to A_{21} and repeat the iteration through the first block row of B. We continue in this fashion until we have completed the first block column of A. In this way, so long as we have chosen the correct block size, the first block row of B will be kept in level 2 cache while the calculations are performed. Then we move to the second block column of A and the second block row of B. We continue with this pattern until the whole matrix has been covered.

The file *matrixMul2.f90* contains the code that performs this calculation as described; the loop indices are in the order corresponding to our description above. To compile this code you will have to include the LAPACK library flag (-llapack) and the BLAS[*] library flag (-lblas) as I have included a call to the LAPACK subroutine 'DGEMM' for comparison to our blocking attempts. Here the block size is left as an adjustable parameter to investigate. This program is less than optimal and I have left it as an exercise for the reader to attempt improvements to the code. You should definitely swap the loop indices to see which order gives the best performance. Another suggested improvement is that you should include code to deal with the situation where the block size does not fit neatly into the matrix. Also, I have simplified matters and used square matrices; more generally they should be rectangular. These improvements *may* or may not include changes to the loop index limits and the use of the Fortran intrinsic function 'MIN'.

---

[*] Basic Linear Algebra Subroutines

Whether or not you implement these improvements, you should find that there is an optimal block size in terms of the performance of the program. This optimum block size will be influenced by the size of the cache levels and cache line for your particular CPU. You should have also noticed that the LAPACK subroutine outperforms our attempts quite spectacularly. This is because when the LAPACK library is installed on to your machine it is optimised to take full advantage of the underlying architecture and the binaries run very efficiently. We can also optimise our code. The GNU Fortran complier that we have been using has many optimisation flags that can be switched on at the command line. They are too numerous to mention here but can be controlled through the use of general optimisation levels using the –O (capital o) flag. There are four general levels of optimisation. The default is level zero (-O0) which switches off all optimisation, this is useful for debugging, developing, and experimenting with code as we are doing in this chapter and throughout this book. Level one (-O1) switches on a number of optimisation options, while level two (-O2) switches on several more. Level three (-O3) switches on all optimisation options associated with the complier. Feel free to experiment with these optimisation flags when compiling your code and see what effects they have on runtime (you'll be surprised). Note that optimisation levels 2 and 3 may have no benefit over level one and may in fact reduce the performance of the binary code. For the rest of this Chapter we'll maintain the default, i.e. no optimisation. For more on the optimisation options go to http://gcc.gnu.org/onlinedocs/gcc/Optimize-Options.html .

There are other ways to improve the speed of matrix multiplications. Strassen's algorithm is one of them. Strassen's algorithm partitions the matrices into sub-matrices then, with a clever bit of manipulation, reduces the number of sub-matrix multiplications required to get the same result by one. Basically, the sub-matrix multiplications are swapped for matrix additions and subtractions. Applying this idea recursively to the sub-matrices we can significantly reduce the required number of operations for matrix multiplication (for large matrices). This recursion can continue until the sub-matrices degenerate into numbers however, in practice it continues until the sub-matrices are of such size that the naïve matrix multiplication algorithm becomes more efficient than continuing the recursion. We already know that the runtime associated with the naïve approach to matrix multiplication is proportional to $O(N^3)$, where N is the size of the matrix (assumed square). Strassen's algorithm provides a runtime that is proportional to $O(N^{2.8074})$. It is worth noting that Strassen's algorithm is less numerically robust that the naïve approach as it involves the subtraction of sub-matrices to compute. If the corresponding elements of those sub-matrices are sufficiently close in value then this will lead to unit round off errors in the result.

12.1.4 Loop Unrolling

Another way of squeezing performance out of a computer program is to unroll incremental loops. That is instead of incrementing the loop index by one on each iteration we increment it by a larger integer value and adjust the contents of the loop appropriately. This increment we call the stride of the loop. For instance, if we were summing the elements of a vector, say, then normally the stride is one and we would write the loop as

$DO\ I\ =\ 1, N$

$\quad SUM\ =\ SUM\ +\ A(I)$

$END\ DO$

However, if we change the stride to two then this changes to

$DO\ I\ =\ 1, N, 2$

$\quad SUM\ =\ SUM\ +\ A(I)\ +\ A(I+1)$

$END\ DO$

Here we increment the array index I by two i.e. $I\ =\ 1, 3, 5, \ldots, N-1$, assuming the vector's size N is an even number. This has the effect of reducing the number of instructions spent controlling the loop, such as pointer arithmetic and testing for the end of the loop. Thus more of time is spent on the calculations we require. It has the additional benefit of hiding inherent latencies, especially the delay in reading variables from memory.

The file *loopRoll.f90* contains code that performs this loop unrolling on the nested loop for matrix addition using stride patterns of one, two and four; note that the matrix dimensions are square and even. After compiling and running this code you should find that you get a significant, albeit modest, performance improvement for the stride pattern of two; on my machine the runtime was reduced by roughly 25%. However, for a stride pattern of four there is little improvement over the stride pattern of two.

Manual loop unrolling is really only advised if we want to squeeze every last bit of performance out of a particular program. As can be seen the process becomes rapidly tedious as we try to extend the unrolling and has diminishing returns in terms of runtime. Some compliers are designed to optimise the binary code produced from your source code and can apply loop unrolling automatically.

12.2 Parallel Programming

Multiple-core machines are now ubiquitous. They offer a means of performing computations in parallel rather than in sequence. For some problems making the computations perform in parallel is rather straightforward, performing a direct numerical integration say, or summing the elements of an array. For other problems making algorithms parallel is not quite such a simple task, for instance matrix factorisations or SOR. The difficulties tend to arise from interdependencies between the different subroutines used to solve the problem or the elements of the array themselves.

No program can run faster than the longest chain of dependent calculations, known in network theory as the critical path. As an analogy consider making a cup of tea. If you want to do this efficiently (i.e. in the quickest time possible) then you would fill and turn on the kettle first. Then as the water boils you would find a mug, put a tea bag in it, fetch the milk, and probably still have time before the water finishes boiling. Once the water has boiled you pour it over the tea bag in the mug, allow it to brew, extract the bag, and add the milk[*]. Note that the critical path here is filling the kettle, waiting for it to boil, adding the hot water, allowing it to brew, extracting the tea bag, and adding the milk. These tasks are dependent on the last and therefore cannot be done in parallel; I dare you to try. Strictly speaking, this is task parallelism rather than a true analogy of multiple-core parallelism; which would be several people making one cup of tea at the same time.

This section does not give an in-depth study of parallel computing but should provide the reader with practical instruction on the use of OpenMP (Multi-Processing) directives within Fortran code, and hopefully draw your interest for further study in the topic.

12.2.1 Many (Hello) Worlds

The OpenMP (OMP) directives do not constitute a new language rather an extension to the FORTRAN we already know. However, just like learning a new language we should still start with a relatively basic program to get used to the syntax. The file *OMP_HelloWorld.f90* contains the program code that will output to screen the text hello world and from which processor the message is coming from. Note that if you have a quad core processor, you should receive four 'hello worlds', an eight core machine you should receive eight hello worlds, and you get the picture. The OMP library should have been installed automatically if the appropriate choice was selected during the Cygwin setup. To compile and run Fortran code using the OMP directives there are two things you must do, first there must be a 'USE OMP_LIB' -- note that this is only necessary to avoid having to declare the OMP functions we are going to use in the program -- statement at the start of the program (which we have seen already using the OMP built in timer). Second we must use the complier flag I mentioned at the start of this chapter.

---

[*] Please, no hate mail if you put the milk in first

Once you have successfully complied and run the program you should have received the appropriate number of hello worlds to your screen, plus the processor identification number the message came from. Note that the processor numbering starts from zero.

If you are unsure of the number of processors on your system compile and run the file *omp_param.f90* which will tell you two things: the maximum number of threads and the maximum number of physical cores on your system. Note that these may be different due to what is called hyper-threading technology on Intel processors. Essentially, hyper-threading allows one physical core to perform two tasks in parallel. Thus an Intel, quad core processor will have a maximum of eight threads. However, if the operating system is unaware of the hyper-threading technology (such as using Cygwin through Windows) then it will see the threads as individual, physical cores. For instance if you have an Intel i7 processor, which is *quad core*, omp_param.f90 may return with the value of eight for both the maximum number of threads available *and* the number of cores on your system. Don't be fooled; you still only have four cores on your system. For AMD processors the maximum number of available threads will be equivalent to the number of cores you processor contains.

Returning to our parallel hello world program let's have a look at the new syntax we've introduced. The 'USE OMP_LIB' statement gives us access to the various precompiled OMP library subroutines and functions (such as 'OMP_GET_NUM_THREADS') without the need to declare their type in our program[*]. The integer variables NTHREADS and TID are used to store the total number of threads on your system and their numeric identification, respectively.

As an aside, if you think of the computational work being done as a line, or thread, on a piece of paper, as the code enters a parallel region the thread separates, or forks, into a number of threads equal to the number of processor cores on your machine (here we ignore the hyper-threading of Intel's processors), performing the computations simultaneously. Each separate thread is being worked

---

[*] If you've done any C++ programming think of it as being like the #include <header_file.h> statements.

on by a separate core and cores do not swap threads (unless specifically programmed to do so). After the parallel region the threads are joined and the work continues on the master thread (or core), which is identified as thread zero. I will attempt to stick to calling them threads for clarity but if I do slip in the odd core here and there you can simply replace it with thread.

Anyhow let's get back to describing the program. The OMP directives are invoked using the syntax '!$OMP' followed by the directive name plus whatever arguments are required. This is necessary as the GNU Fortran compiler does not understand how to interpret parallel regions of code and so these must be handled by OMP specifically; hence the need for the compilation flag at the command line. You can think of the OMP directives as add-ons to the original Fortran language. Each thread prints to screen the hello message and its identification number, with the master thread (TID = 0) printing out the total number of threads. The parallel region is closed using the familiar END statement; like most Fortran syntax think about it as closing an open bracket.

By default all variables scoped in a parallel region are shared by each thread; that is each thread has access to a global copy of the variable. The only exception to this is the iteration index of the first loop after the OMP 'DO' directive, where each thread gets its own private copy by necessity; we shall describe this directive in more detail shortly. It is best practice to explicitly scope all the variables as either shared or private as we enter a parallel region to ensure the code is clear to read and performs as required. Just to point out the nomenclature of OMP: at the top level there are *directives*; directives have *clauses*; clauses may or may not take arguments. For instance, the *parallel directive* has a *shared clause*, which takes the variables of the program as arguments.

If you're ever unsure that a parallel region of code is actually parallel, i.e. making use of all the processors on your machine, then enter the 'Task Manager' (or equivalent if you're not on a Windows machine) and select the 'Performance' tab to see a visual display of each processor. If the parallel regions are working correctly you should see that all the processors work at 100% (you may need to add a large loop somewhere to create sufficient runtime to see this).

You will have noticed that due to the exclamation mark (!) Emacs treats the OMP directive statement as comment, making it a little difficult to distinguish from the other comments in the code. I would recommend leaving blank lines around the directives to make it easier to read. Note that in free-form Fortran the directive can start in any column of the editor.

Now that we have seen some of the syntax of OpenMP let's apply that to something more mathematical.

12.2.2 Vector Summation

Let's start with the relatively simple task of summing the elements of a vector; this has uses within statistics for finding the mean value of a data set, say. On a single core serial processor we would loop through the entire array incrementing the index by one and updating the sum. For a parallel architecture we can split the entire vector into equally sized chunks and perform the summation on each chunk simultaneously. To obtain the sum for the entire vector we would then add up the contributions from each chunk on a single thread. If we define T_S as the time taken to perform the summation on a single core, in a serial manner, then we might expect the time taken for the code to run in parallel to be T_S/P, where P is the number of threads used. This is known as the ideal case; the speedup in program runtime is directly proportional to the number of threads used. In general the speedup is given by the ratio T_S/T_P, where T_P is the time taken to run the code in parallel on P threads.

For the sake of rigour the serial time T_S is *not* the same as T_1. That is, T_S is the time taken to perform the code as normally written, whereas T_1 is the time taken to perform the code where we have spawned a parallel region using only one thread. Typically, $T_1 > T_s$ by a small yet significant amount. That said sometimes the "serial" code may be threaded by the operating system automatically meaning that our speedup measurements might be skewed if we assume the computations are being carried out on a single core. So long as we are clear about which measurement of the "serial" time we are taking we shouldn't run into problems. For the rest of this chapter we will take the serial time as T_1. Also note that an even more rigorous treatment of speedup is to consider the number of floating point operations per second (flops) rather than runtime as the flops measure is independent of the computer architecture used. For the purposes of our discussion we will stick with runtime as a suitable measure of performance.

The file *vectorSum.f90* contains code to perform the vector summation in parallel. You will see in the code that we have introduced the parallel 'do' loop directive (!$OMP DO). This directive must be immediately followed by the iteration loop we wish to make parallel, and close by the corresponding parallel end do directive (!$OMP END DO). In its current, default state the directive will divide the iterations among threads equally such that thread zero gets the chunk defined by $I = 1, ..., N/P$, thread one gets the chunk defined by $I = 1 + N/P, ..., 2N/P$, and so on; where the size of the loop is not exactly divisible by the number of threads the remainder gets evenly distributed. This is called static assignment. We can change this assignment through use of a scheduling clause. The syntax for this clause is relatively simple and all we do is type '*schedule (type, chunk)*' after the OMP do directive. Here type is either '*static*', '*dynamic*', '*guided*' or '*auto*', and *chunk* is an integer value that has slightly different meanings depending on the type used.

We have already seen that *static* assignment allocates all the iterations to each thread before they execute at runtime, with the assignment being divided equally amongst the threads by default. We can change this behaviour by specifying a chunk size. For instance, if we set chunk equal to one then the assignment of then iterations $I = 1, ..., P$ are assigned to threads 0 to $P - 1$ respectively, then iterations $I = P + 1, ..., 2P$ are assigned to threads 0 to $P - 1$ respectively, and so on. In *dynamic* assignment only some of the iterations are assigned to threads before execution of the loop; the number being set by chunk. Once a particular thread finishes its allotted iteration(s), it requests another chunk of iterations from those that remain. If you were to print out which thread did which iterations you would find that the assignment would be (somewhat) random from execution to execution. *Guided* allocates a large portion of iterations to each thread dynamically, as above, but then decreases the portion size after each successive allocation until it reaches a minimum size specified by chunk. *Auto* allows the complier to decide the best type and chunk to use at runtime.

The overall scheduling strategy is to ensure none of the threads are idle during the parallel sections and that we make the most effective use of our parallel machine. Essentially, we are insuring a uniform distribution of work across all threads. Feel free to play around with the scheduling clause to see if it affects the performance of the program by any significant amount.

The *reduction* clause allows the variable 'SUM' to be updated as the addition of each thread's local value of 'SUM' on the master thread. To clarify, each thread *reads* the value of 'SUM' from the shared memory, creating a local copy on which to work. *Without* the reduction clause once the work has been completed each thread writes it's value of 'SUM' back to the shared location. This means that the updated value of 'SUM' would be whichever thread finished last. *With* the reduction clause each thread, once finished, passes its local value of 'SUM' to the master thread which then combines them as specified by the operator argument in the clause; in this case it adds them ('REDUCTION (+:SUM)').

Compile and run *vectorSum.f90*. Did you obtain the speedup you expected? Was it anywhere near the ideal case? On an AMD, quad core processor I found the speedup (T_S/T_4) to be around 3.7. On an Intel, quad core processor using hyper-threading technology running the same code, I found that the best observed speedup (T_S/T_8) was around 5.1. (Remember that both these processors have four physical cores thus hyper-threading does offer a performance benefit, but not as much as actually having eight physical cores).

Note that to obtain statistically valid timing results the summations are repeated 100 times for both the serial and parallel regions. This is easily achieved for the serial loop as we simply include an outer conditional 'do while' loop. However, doing the same for the parallel region is a little unfair as the code has to spawn or fork team of threads at the beginning of each 'do while' loop, then destroy or join them at the end of each loop. This requires a small but significant amount of time to process. This amount of time is called an overhead as it is the price we (always) pay to perform parallel execution of code. How might you make the comparison more fair yet still obtain statistically valid timing results?

12.2.3 Overheads: Amdahl vs. Gustafson

Generally, we find that the improvement in performance due to spawning a parallel region of code is limited by overheads. Overheads arise from a number of places not least the portions of the code that are necessarily serial, and the communication (or message passing) between threads. For instance, in our vector summation example the reduction at the end of the procedure to obtain the total for the entire vector has to be done in serial on the master thread. For all parallel regions of code there is one overhead that is unavoidable; the time it takes fork (create) and join (destroy) parallel threads.

Obviously the point of increasing the number of threads used in a parallel region of code is to improve the performance (runtime or flops) of that region. A typical program has regions that are both parallel and serial. As we increase the number of threads, the parallel regions compute faster but the serial portions are unaffected. Thus the overall performance of the program is limit by the amount of time it takes to compute the serial regions of code. This is Amdahl's Law. We can formulate this law into an equation which states that the speedup in a program is given by

$$S(P) \equiv \frac{T_S}{T_P} = \frac{T_S}{T_S\left[(1-\beta)+\beta/P\right]} = \frac{1}{(1-\beta)+\frac{\beta}{P}} \qquad (12.2)$$

where β is the portion or fraction of the code that is or can be made parallel. We can see this formulation makes sense in that with no serial fraction of code (β=1) the speedup is simply given by the number of threads used in the parallel region (ideal case). Obvious but worth pointing out that when the code is completely serial (β=0) then there is no speedup i.e. S(P)=1. Figure 12.2 plots the consequence of Amdahl's Law on the speedup of various programs compared to the number of processors used. Each line represents a program with different fractions of code that are able to be made parallel and shown for comparison is the ideal case (β=1). These results seem quite pessimistic, where even a code with only a 10% serial portion diverges significantly from the ideal case for a relatively low number of processors. Eventually, for any code less than ideal adding more processors fails to further improve the performance. In reality, the situation is worse as we have not considered the overhead due to communication between processors, and adding more processors may actually decrease the speedup after a certain point.

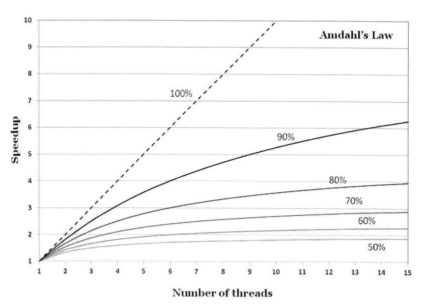

Figure12.2: Amdahl's Law for the speedup of programs as a function of the number of threads used.

We can test Amdahl's Law by changing the number of threads we use in the parallel region of code. OMP comes with a suite of library subroutines that can affect hardware parameters. For instance, we can set the number of threads available to the program (up to the maximum number of threads we have on our system) by specifying that number as the argument to the subroutine call 'OMP_SET_NUM_THREADS'. In this way, we can see how much speedup we obtain using a different number of threads in the parallel region.

Before we throw our toys out if the pram and claim that parallel programming is fundamentally flawed it must be noted that Amdahl's Law treats the problem as having a fixed amount of work to do, and measuring the time taken to do that work. Gustafson argues that programmers tend to set the size of problems to use the available equipment to solve those problems within a practical fixed time. Hence, if faster, i.e. more parallel, machines are available, larger problems can be solved in the same amount of time as smaller problems on slower, less parallel, machines. In essence, we can think of the individual processor work-load as remaining fixed and as we add more processors we necessarily solve a bigger problem. The formula for Gustafson's Law is given by

$$S(P) = P - \alpha(P-1) = \beta P + \alpha \tag{12.3}$$

where α is the fraction of the code that cannot be made parallel i.e. the serial portion of the program ($\alpha + \beta = 1$). It should be noted that in the formulations of Amdahl's Law and Gustafson's Law it is assumed that the parallel portions of code are uniformly distributed amongst all P threads.

If we go back to our tea making analogy of parallelism at the start of this section I mentioned that multi-core parallelism is like several people making one cup of tea at the same time, with only one kettle, one teabag, one cup and so on. This is actually Amdahl's view. In Gustafson's view each person makes one cup of tea where there is enough equipment i.e. kettles, teabags, cups etc. for each person.

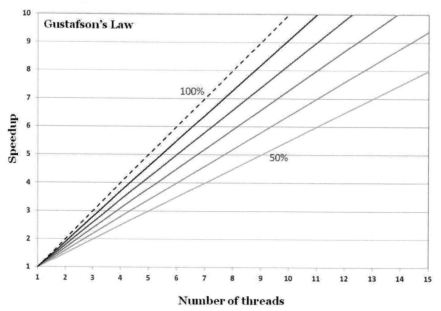

Figure 12.3: Gustafson's Law for the (scaled) speedup of programs as a function of the number of threads used.

We can test Gustafson's Law by increasing the size of our summation vector for a fixed number of threads; this should vary α such that we should be able to quantify its value to some degree of accuracy for some given vector size.

In either case of Amdahl's Law or Gustafson's Law it is clearly beneficial to make $\alpha = (1 - \beta)$, i.e. the strictly serial portion of the code, as small as possible.

12.3 More Stuff to Know...

Sometimes when running parallel threads we wish to control their behaviour in some manner. As it stands we simply spawn the threads, let them go, and hope that the scheduling applied by the complier keeps them from being idle. There are, however, ways of having greater control over thread behaviour. For instance, by default there is a 'WAIT' command applied at the end of a parallel do loop; this is referred to as a barrier. This means that any thread must wait at the end of the do loop until all threads have completed executing the loop before moving on; this is called thread synchronisation. We can change this by explicitly writing the 'NOWAIT' clause after the OMP end do directive. This allows any thread that has completed the loop to immediately continue with the next portion of code i.e. it removes the implied barrier.

Other synchronisation constructs exist within the OMP framework and can be used to give the programmer greater control over the parallel threads. There are several of these constructs but of the most important of these are the 'master', 'critical', 'barrier' and 'atomic' directives. For more information on these directives and OpenMP in general please visit https://computing.llnl.gov/tutorials/openMP/; I have found this reference to be extremely useful.

Exercises

1. Write a program to determine the size of the level 1 and level 2 caches on your machine (and level 3 if your processors has it).

2. Write a program to determine the cache line size on your machine.

3. Check that the optimal index order for the naive matrix multiplication algorithm is (i, j, k) for Fortran. Can you explain why that is/is not the case?

4. Apply loop unrolling to the outer loop of the matrix multiplication algorithm. Does this improve the performance of the multiplication and why/why not?

5. Write a program that performs matrix multiplication in parallel for different numbers of threads. Determine the value of α (or β) in your code for a given matrix size using Amdahl's law. What is causing this serial portion of code and can you quantify the amount of runtime it takes?

6. Using your value of α from the previous question does Gustafson's Law (scalable speedup) hold true?

7. Choose any numerical quadrature we have discussed in previous chapter and attempt to write it in parallel code. Why is writing parallel code for an initial value, ODE problem fundamentally flawed?

8. The Jacobi iteration scheme to approximately solve a general second order ODE is given by

$$f_i^{(n)} = -\frac{1}{\theta}\left(\varphi f_{i+1}^{(n-1)} + \psi f_{i-1}^{(n-1)} - \delta x_i\right)$$

where θ, φ, and ψ are constants related to the coefficients of the differential equation; i is the index of the discrete grid approximation of the continuous space x; f is some quantity in that space; and n is the iteration count. Write a parallel program that exploits the "red-black" nature of this scheme. (Tip: Use the program(s) from Chapter 9 as a guide)

9. Attempt to do the same for the Gauss-Seidel iteration scheme (or SOR). Encounter any difficulties?

Bibliography

The following is a list of reference literature that I found very useful in writing this book and the programs contained within. I also provide a guide to more general reading to the topics covered in this text.

Computational Physics

DeVries, P. L., *A First Course in Computational Physics*, John Wiley & Sons, 1994.

Klein, A. & Godunov, A., *Introductory Computational Physics*, Cambridge University Press, 2006.

Koonin, S. E. & Meredith, D. C., *Computational Physics: Fortran Version*, Addison-Wesley, 1990.

Landau, R. H., Páez, M. J., & Bordeianu, C. C., *A Survey of Computational Physics: Introductory Computational Science*, Princeton University Press, 2008.

Pang, T., *An Introduction to Computational Physics,* 2nd ed., Cambridge University Press, 2006.[*]

---

[*] This is the Java version. The 1st edition, published in 1997, is the Fortran version.

General reading ...

Classical Mechanics

Goldstein, H., Poole, C. P., & Safko, J. L., *Classical Mechanics*, 3rd ed., Addison Wesley, 2001.

Kleppner, D. & Kolenkov, R. J., *An Introduction to Mechanics*, McGraw Hill, 1973.

McCall, M. W., *Classical Mechanics*, John Wiley and Sons, 2001.

Finite Element Method

Braess, D., *Finite Elements*, Cambridge University Press, 2001.

Johnson, C., *Numerical Solution of Partial Differential Equations by the Finite Element Method*, Cambridge, 1987.

Fourier Analysis

Briggs, W. L. & Henson, V. E, *The DFT: An Owner's Manual for the Discrete Fourier Transform*, SIAM 1995.

Dym, H. & McKean, H. P., *Fourier Series and Integrals*, Academic Press, 1972.

Folland, G. B., *Fourier Analysis and Its Applications*, Brooks/Cole Publishing Co., 1992.

Körner, T. W., *Fourier Analysis*, Cambridge University Press, 1988.

Tolstov, G. P., *Fourier Series*, Dover, 1972.

Walker, J. S., *Fourier Analysis*, Oxford University Press, 1988.

General Physics

Orzel, C., *How to Teach Physics to Your Dog*, Scribner, 2009.

Susskind, L. & Hrabovsky, G., *The Theoretical Minimum: What You Need to Know to Start Doing Physics*, Allen Lane, 2013.

High Performance/ Parallel Computing

Chandra, R., Dagum, L., Kohr, D., Maydan, D., McDonald, J. & Menon, R., *Parallel Programming in OpenMP*, Academic Press, 2000.

Dowd, K. & Severance C., *High Performance Computing*, 2nd ed., O'Reilly, 1998.

Hager, G. & Wellein, G., *Introduction to High Performance Computing for Scientists and Engineers*, CRC Press, 2010.

Koelbel, C. H., *The High Performance Fortran Handbook*, MIT Press, 1994.

Interpolation & Approximation

Davies, P. J., *Interpolation and Approximation*, Blaisdell, 1963, (reprinted by Dover, 1975).

Nürnberger, G. , *Approximation by Spline Functions*, Springer, 1989.

Powell, M.J.D., *Approximation Theory and Methods*, Cambridge, 1981.

Monte Carlo Method

Sobol, I. M., *A Primer for the Monte Carlo Method*, CRC Press, 1994.

Numerical Analysis

Demmel, J.W., *Applied Numerical Linear Algebra*, SIAM, 1997.

Golub, G. H. & Van Loan, C., *Matrix Computations*, 3rd ed., Johns Hopkins, 1996.

Higham, N. J., *Accuracy and Stability of Numerical Algorithms*, SIAM, 1996.

Overton, M. J., *Numerical Computing and the IEEE Floating Point Standard*, SIAM, 2001.

Ordinary Differential Equations

Hairer, E., Norsett, S. P. & Wanner, G., *Solving Ordinary Differential Equations I: Nonstiff Problems*, 2nd ed., Springer, 2000.

Hairer, E. & Wanner, G., *Solving Ordinary Differential Equations II: Stiff and Differential-Algebraic Problems*, 2nd ed., Springer, 2004.

Lambert, J. D., *Numerical Methods for Ordinary Differential Equations: The Initial Value Problem*, 2nd ed., John Wiley and Sons, 1991.

Shampine, L. F., *Numerical Solution of Ordinary Differential Equations*, Chapman & Hall, 1994.

Partial Differential Equations

Folland, G., *Introduction to Partial Differential Equations*, 2nd ed., Princeton, 1995.

John, F. , *Partial Differential Equations*, 4th ed., Springer, 1991.

Richtmeyer, R.D. & Morton, K. W., *Difference Methods for Initial-Value Problems*, 2nd ed., Krieger, 1994.

Taylor, M. E., *Partial Differential equations: Basic Theory*, Springer, 1996.

Quantum Mechanics

Gasiorowicz, S., *Quantum Physics*, John Wiley and Sons, 1974.

Griffiths, D. J., *Introduction to Quantum Mechanics*, Prentice Hall, Englewood Cliffs, 1995.

Landau, L. D. & Lifshitz, E. M., *Quantum mechanics: non-relativistic theory*, 3rd ed., Pergamon Press, 1991.

Rae, A. I. M., *Quantum Mechanics*, 3rd ed., Institute of Physics, London, 1998.

Appendix

Here we list all the programs referred to in the text in the order that they appear. For convenience, you can also find the code at the following web address:

http://compphysintro.wordpress.com/computational-physics-an-undergraduates-guide/

Chapter 2 – Getting Comfortable

PiPolygon.f90

```fortran
PROGRAM PI_POLYGON

 IMPLICIT NONE

 INTEGER, PARAMETER :: N = 4

 INTEGER, PARAMETER :: NRHS = 1

 INTEGER I, J

 INTEGER INFO, IPIV(N)

 DOUBLE PRECISION A(N,N), P

 DOUBLE PRECISION B(N)

 EXTERNAL DGESV

 DO J=1,N

   DO I=1,N

     P = (I+2)*(J-1)

     A(I,J) = 1/(2**P)

   END DO

 END DO

 PRINT *, 'Coefficient matrix A:'

 !

 DO I=1,N
```

```
      WRITE(*,100) A(I,1), A(I,2), A(I,3), A(I,4)
100 FORMAT(4(' ', f10.6))
    END DO
    !
    B(1) = 3.061467
    B(2) = 3.121445
    B(3) = 3.136548
    B(4) = 3.140331
    !
    CALL DGESV(N, NRHS, A, N, IPIV, B, N, INFO)
      PRINT *, 'Display the solution x'
      DO I = 1,N
        WRITE(*,101) B(I)
101 FORMAT(1(' ', f10.6))
      END DO
      STOP
END PROGRAM PI_POLYGON
!END OF FILE**************************************************
```

Chapter 3 – Interpolation and Data Fitting

linearInterp.f90

```
!*****************************************************************
! Program to calculate the linear interpolation of the function
! f(x) = sinc(x^2) using 10 points as simulated measured data.
! Data is saved to a text file for graphical use elsewhere --
! gnuplot, Excel, MATLAB,...
!*****************************************************************
PROGRAM LINEARINTERP
```

Appendix

```
IMPLICIT NONE

! Line spaces represented by arrays, pts defines the number of
! points to use.
INTEGER,      PARAMETER :: PTS = 100
INTEGER,      PARAMETER :: DPTS = 10
DOUBLE PRECISION, PARAMETER :: INTERVAL = 5.D0
! Declare the variables you are going to use.
! single dimension arrays of size pts
DOUBLE PRECISION, DIMENSION(PTS)    :: X,G1, G2
DOUBLE PRECISION, DIMENSION(DPTS, 2) :: D
DOUBLE PRECISION F, ALPHA, P
INTEGER I, K, M
EXTERNAL F

ALPHA = INTERVAL/PTS
M = (PTS - 1)/(DPTS -1)

DO I = 1, PTS
   P = I        ! Assignment converts int into real
   X(I) = P*ALPHA  ! Set up the x values which are floats.
END DO

! Assign the data points we are taking as "measurements"
! Here we use 10 equidistant points
DO K = 1, DPTS
  D(K,1) = X((K-1)*M +1)
  D(K,2) = F( D(K,1) )
END DO

CALL LINEAR(X, D, G1, G2, M, PTS, DPTS)

! Write the data to a text file.
! fw.d float total-width . decimal-range
! The output will be space delimited using 2X
```

```fortran
OPEN(UNIT=1, FILE="linear_interp_data.txt")

WRITE(1,100) (X(I), F(X(I)), G1(I), G2(I), I=1,PTS)

100 FORMAT(4(f16.10,2X))

CLOSE(1) ! File must be closed prior to exiting

END PROGRAM
```

!***

```fortran
DOUBLE PRECISION FUNCTION F(X)

  DOUBLE PRECISION X

  F = SIN(X*X)/(X*X)

END FUNCTION
```

!***

```fortran
SUBROUTINE LINEAR(X, D, G1, G2, M, PTS, DPTS)

  ! Declare the variables you are going to use.

  ! single dimension arrays of size pts

  INTEGER I, J, K, M, PTS, DPTS

  DOUBLE PRECISION, DIMENSION(PTS)   :: X, G1, G2

  DOUBLE PRECISION, DIMENSION(DPTS,2) :: D

  DOUBLE PRECISION C1, C2

  ! Assign the data points we are taking as "measurements"

  ! to our approximation function g(x)

  DO K = 1, DPTS

    G1((K-1)*M +1) = D(K, 2)

    G2((K-1)*M +1) = D(K, 2)

  END DO

  ! Calculate the linear interpolation. Max J is reduced to dpts-1 as

  ! index I would go past last "measured" data point otherwise.

  ! The values for I select the x and f(x) between "measured" points

  DO J = 1, DPTS-1

    DO I = (J-1)*M+2, M*J
```

```
        ! Nested loop. Here j remains constant while i iterates
        ! d[j,1], d[j+1,1], d[j,2], and d[j+1,2] are our
        ! x and f(x) "measurement" values, respectively
        C1 = ( X(I) - D(J+1,1) )/( D(J,1)- D(J+1,1) )
        C2 = ( X(I) - D(J,1) ) /( D(J+1,1) - D(J,1) )
        G1(I) = D(J,2) + C2*(D(J+1,2) - D(J,2)) ! non-symmetrical form
        G2(I) = C1*D(J,2) + C2*D(J+1,2)         ! symmetrical form
      END DO
    END DO
    RETURN
  END SUBROUTINE
!END OF FILE ****************************************************
```

lagrangeInterp.f90

```
!****************************************************************
! Program to calculate the Lagrange interpolation of the function
! f(x) = sinc(x^2) using 10 points as simulated measured data.
! Polynomials of order n = 1 to 7 are calculated.
! Data is saved to a text file for visuals elsewhere,
! e.g. gnuplot, Excel, MATLAB,...
!****************************************************************
PROGRAM LAGRANGEINTERP
  IMPLICIT NONE
  ! Line spaces represented by arrays defined by following parameters
  INTEGER,        PARAMETER :: PTS = 100
  INTEGER,        PARAMETER :: DPTS = 10
  DOUBLE PRECISION, PARAMETER :: INTERVAL = 5.0
  INTEGER,        PARAMETER :: N_MAX = 9
  ! Declare the variables you are going to use.
  DOUBLE PRECISION, DIMENSION(PTS)       :: X
```

```fortran
DOUBLE PRECISION, DIMENSION(PTS, N_MAX)  :: G

DOUBLE PRECISION, DIMENSION(DPTS, 2)    :: D

INTEGER M           ! Multiplier to obtain equidistant points

INTEGER I,J,K,N     ! Loop index integers

DOUBLE PRECISION ALPHA, P, F

EXTERNAL F

ALPHA = INTERVAL/PTS

M = (PTS - 1)/(DPTS -1)

DO I = 1,PTS

   P = I

   X(I) = P*ALPHA ! Set up the x values which are floats.

END DO


! Assign the data points we are taking as "measurements"

! Here we use 10 equidistant points

DO K = 1, DPTS

   D(K,1) = X((K-1)*M +1)

   D(K,2) = F( D(K,1) )

END DO

! Initialise the approximating functions to zero

DO J = 1, N_MAX

   DO I = 1, PTS

      G(I,J) = 0.0

   END DO

END DO

! Loop over polynomial orders n=1 to n=N_MAX=7

DO N = 1, N_MAX

   CALL LAGRANGE(X, D, G, M, PTS, DPTS, N)

END DO
```

Appendix

```
! Write the data to a text file.
! The output will be space delimited columns.
OPEN(UNIT=1,FILE="lagrange_interp_data.txt")
WRITE(1,100)( X(I), F(X(I)), G(I,2), G(I,4), &
   G(I,6), G(I,8), G(I,9), I = 1, PTS)
100 FORMAT(7(2X,F16.10))
  CLOSE(1) ! File must be closed prior to exiting
END PROGRAM
!*********************************************************************
SUBROUTINE LAGRANGE(X, D, G, M, PTS, DPTS, N)
  ! Declare the variables you are going to use.
  INTEGER PTS, DPTS, N, M
  DOUBLE PRECISION, DIMENSION(PTS)     :: X
  DOUBLE PRECISION, DIMENSION(PTS, N)  :: G
  DOUBLE PRECISION, DIMENSION(DPTS, 2) :: D
  DOUBLE PRECISION, DIMENSION(DPTS, PTS):: LAM
  INTEGER I,J,K,L
  ! Initialise the Lagrange coefficients to one
  DO J = 1,PTS
    DO I = 1, DPTS
      LAM(I,J) = 1.0
    END DO
  END DO
  !Loop for the Lagrange interpolation
  DO J = 1, DPTS-N, N          ! intervals; stride = n
    DO I = (J-1)*M+2, M*(J-1+N)+1 ! values between intervals
      DO K = J, J+N             ! data points defining interval
        DO L = J, J+N           ! loop to determine coefficients
          ! ensure we miss the case when k == l
```

```fortran
      IF(K.NE.L)THEN
        LAM(K,I) = LAM(K,I)*(X(I)-D(L,1))/(D(K,1)-D(L,1))
      END IF
    END DO ! loop l
    ! Calculate the approximation.
    G(I,N) = G(I,N) + LAM(K,I)*D(K,2)
   END DO ! loop k
  END DO ! loop i
 END DO ! loop j
 RETURN
END SUBROUTINE
!******************************************************************
DOUBLE PRECISION FUNCTION F(X)
 DOUBLE PRECISION X
 F = SIN(X*X)/(X*X)
END FUNCTION
!END OF FILE******************************************************
```

linearLeastSquares.f90

```fortran
!******************************************************************
!  Program to compute the least linear squares data fit to the data
!  stored in 'llsdata.txt'
!******************************************************************
PROGRAM LLS
 IMPLICIT NONE
 DOUBLE PRECISION, ALLOCATABLE :: A(:,:), B(:), XI(:), YI(:), T(:)
 INTEGER I,J,K,N,M,INFO
 INTEGER, ALLOCATABLE :: IPIV(:)
 ! M - order of the approximating polynomial
```

Appendix

```
!    i.e. y(x) ~ a0 + a1*x + ... + am*x^m
! N - total number of data point pairs
! INFO and IPIV are required for the LAPACK routine
M = 2
ALLOCATE ( A(M+1,M+1), B(M+1), IPIV(M+1) )
A = 0.0D0
B = 0.0D0
OPEN(UNIT=1, FILE="llsdata.txt")
READ(1,*) N ! N saved in first line of text file
ALLOCATE ( XI(N), YI(N), T(N) )
T = 1.0D0
I = 1
DO WHILE(1.EQ.1)
    READ(1,*, END=10) XI(I), YI(I)
    ! x & y data saved in two separated columns
    I = I + 1
END DO
10 CONTINUE
CLOSE(1)
! Error check total number of data point pairs match
IF((I-1).NE.N) THEN
    PRINT *, 'Number of data points does not match header'
    STOP
END IF
! Set up matrix A and RHS vector B
DO J = 1,M+1
    DO K = 1,N
        A(1,J) = A(1,J) + T(K)
        A(2,J) = A(2,J) + T(K)*XI(K)
```

```
      A(3,J) = A(3,J) + T(K)*XI(K)*XI(K)

      B(J) = B(J) + YI(K)*T(K)

    END DO

    DO I = 1,N

      T(I) = T(I)*XI(I)

    END DO

  END DO

! Solve the linear system of equations using LAPACK's DGESV routine

  CALL DGESV(M+1, 1, A, M+1, IPIV, B, M+1,INFO)

  WRITE(6,*) 'X = '

  WRITE(6,101) (B(I), I=1,M+1)

101 FORMAT (F10.6)

   DEALLOCATE (A,B,XI,YI, T, IPIV)

END PROGRAM LLS

!END OF FILE ***********************************************
```

Contents of *llsdata.txt*

4

1.0 2.0

2.0 1.0

3.0 3.0

4.0 6.0

Appendix

Chapter 4 – Searching for Roots

bisection.f90

!***

! Program performs root finding by the bisection method

!***

PROGRAM BISECTION

  IMPLICIT NONE

! Declarations

  DOUBLE PRECISION A, B, R, F

! A -- initial interval value

! B -- final interval value

! R -- computed root

! F -- function to be evaluated

! External means F defined in separate subprogram unit

  EXTERNAL F

! Initialise variables

  A = 0.D0

  B = 1.57D0

! Call subroutine to perform bisection

  CALL BISECT(A, B, R, F)

! print results to screen or to file

  PRINT *, 'Root of f(x) = cos(x) - x is ', R

END PROGRAM BISECTION

!***

DOUBLE PRECISION FUNCTION F(X)

  DOUBLE PRECISION X

  F = COS(X) - X

END FUNCTION F

```
!************************************************

SUBROUTINE BISECT( LEFT, RIGHT, MID, F)

! Declare variables
  DOUBLE PRECISION LEFT, RIGHT, MID, F
  DOUBLE PRECISION FL, FR, FM, ERROR
  DOUBLE PRECISION, PARAMETER :: TOL = 1.D-8
  INTEGER I

! Initialise variables
  I = 0
  ERROR = 1.D0
  FL = F(LEFT)
  FR = F(RIGHT)

! Check that interval brackets root
  IF(FL*FR.GT.0)THEN ! i.e. either both +ve or both -ve
     MID = -9999 ! unlikely the root will be -9999
     RETURN
  END IF

  DO WHILE (ERROR.GT.TOL)
     I = I + 1
     MID = (LEFT + RIGHT)/2
     FM = F(MID)
     IF(FL*FM.LT.0)THEN ! Root bracketed in left interval
        RIGHT = MID
        FR = FM
     ELSE ! Root bracketed in the right interval
        LEFT = MID
        FL = FM
     END IF
```

Appendix

```
     ERROR = ABS( RIGHT-LEFT )
   END DO
   PRINT *, 'No of iters:', I
   RETURN
 END SUBROUTINE BISECT
! END OF FILE ********************************************
```

NewtRaph.f90

```
!********************************************************
! Program to compute the roots of a function using the
! Newton-Raphson method
!********************************************************
PROGRAM NRROOTS
  IMPLICIT NONE
  DOUBLE PRECISION X, F, DF
  EXTERNAL F, DF
  X = 0.8D0
  CALL NEWTRAPH( X, F, DF )
  PRINT *, 'Root of f(x) = cos(x)-x is ', X
END PROGRAM NRROOTS
!********************************************************
DOUBLE PRECISION FUNCTION F(X)
  DOUBLE PRECISION X
  F = COS(X) - X
END FUNCTION F
!********************************************************
DOUBLE PRECISION FUNCTION DF(X)
  DOUBLE PRECISION X
```

```fortran
      DF = -SIN(X) - 1.D0
    END FUNCTION DF

!**********************************************
    SUBROUTINE NEWTRAPH( X, F, DF )
      DOUBLE PRECISION X, F, DF, DELTA, ERROR
      DOUBLE PRECISION, PARAMETER :: TOL = 1.D-08
      INTEGER I
      DO WHILE (ERROR.GT.TOL)
        I = I + 1
        DELTA = -F(X)/DF(X)
        X = X + DELTA
        ERROR = ABS( DELTA )
      END DO
      RETURN
    END SUBROUTINE
! END OF FILE ****************************************
```

NRBHybrid.f90

```fortran
!****************************************************
! Program that implements a hybrid Newton-Raphson
! Bisection method to find the roots of a function.
! The root is known to lie within an interval [A,B]
! and the result is returned to R. If the NR step
! lies with the bounds then it is accepted, else a
! bisection step is taken.
!****************************************************
    PROGRAM NRBHYBRID
      IMPLICIT NONE
```

Appendix

```
DOUBLE PRECISION A, B, R, F, DF

EXTERNAL F, DF

A = 0.3D0

B = 0.7D0

CALL NRBISEC( A, B, R, F, DF )

PRINT *, 'Root at xr =',R

PRINT *, 'f(xr) =',F(R)

END PROGRAM NRBHYBRID

!****************************************************

DOUBLE PRECISION FUNCTION F(X)

  DOUBLE PRECISION X, X8, X6, X4, X2

  DOUBLE PRECISION, PARAMETER :: C8 = 6435, C6 = -12012

  DOUBLE PRECISION, PARAMETER :: C4 = 6930, C2 = -1260

  DOUBLE PRECISION, PARAMETER :: C0 = 35, D = 128

  X8 = X*X*X*X*X*X*X*X

  X6 = X*X*X*X*X*X

  X4 = X*X*X*X

  X2 = X*X

  F = (C8*X8 + C6*X6 + C4*X4 + C2*X2 + C0)/D

END FUNCTION F

!****************************************************

DOUBLE PRECISION FUNCTION DF(X)

  DOUBLE PRECISION X, X7, X5, X3, X1

  DOUBLE PRECISION, PARAMETER :: C8 = 6435, C6 = -12012

  DOUBLE PRECISION, PARAMETER :: C4 = 6930, C2 = -1260

  DOUBLE PRECISION, PARAMETER :: D = 128

  X7 = X*X*X*X*X*X*X

  X5 = X*X*X*X*X

  X3 = X*X*X
```

```
        X1 = X

        DF = (8*C8*X7 + 6*C6*X5 + 4*C4*X3 + 2*C2*X1)/D

    END FUNCTION DF

!*******************************************************

SUBROUTINE NRBISEC( A, B, R, F, DF )

    DOUBLE PRECISION A, B, R, F, DF

    DOUBLE PRECISION FA, FB, FR, DFR, DELTA, ERROR

    INTEGER I

    DOUBLE PRECISION, PARAMETER :: TOL = 5.D-03

    ! initialise variables

    FA = F(A)

    FB = F(B)

    ERROR = 1.D0

    I = 0

    IF(FA * FB .GT. 0)STOP 'Root not bracketed'

    ! Select a "best" starting point - the limit with the

    ! smallest function value is likely closest to the root

    IF( ABS(FA) .LE. ABS(FB) ) THEN

        R = A

        FR = FA

    ELSE

        R = B

        FR = FB

    END IF

    DFR = DF(R)

    ! Set up the while loop to exit when the tolerance met

    DO WHILE( ERROR .GT. TOL )

    ! if iteration counter goes above 30 stop execution

        I = I + 1
```

Appendix

```
    IF(I.GT.30)STOP 'Root not converged after 30 iters'
! Decide if using NR or Bisection for this step
    IF( (DFR * (R - A) - FR)&
        *(DFR * (R - B) - FR).LE.0 )THEN ! NR
      PRINT *, 'NR'
      DELTA = -FR/DFR
      R = R + DELTA
    ELSE ! bisection
      PRINT *, 'BI'
      DELTA = (B - A)/2.D0
      R = (A + B)/2.D0
    END IF
    FR = F(R)
    DFR = DF(R)
    IF(FA * FR .LE. 0)THEN
      B = R
      FB = FR
    ELSE
      A = R
      FA = FR
    END IF
    ERROR = ABS(DELTA/R)
  END DO
  PRINT*, 'No of iters:', I
  RETURN
END SUBROUTINE NRBISEC
! END OF FILE ****************************************
```

bruteSearch.f90

```fortran
!*******************************************************
SUBROUTINE SEARCH ( A, B, STEP, MAXVAL, N, M, F )
! Searches for the roots of the function F on the
! interval [A,B] using STEP size. There should be
! N roots in this interval. M found roots returned.
    DOUBLE PRECISION, DIMENSION (N) :: A, B
    DOUBLE PRECISION STEP, FA, FB, MAXVAL
    INTEGER I, RCOUNT, M
! Min value store in A(1)
! Evaluate function at minval for initial check
    FA = F(A(1))
    M = N
    RCOUNT = 0
    I = 1
! Terminate loop if the search reaches the end of the interval
    DO WHILE ( B(I).LT.MAXVAL )
! Move the search forward one step
        B(I) = A(I) + STEP
! Evaluate the function at new position
        FB = F( B(I) )
! Check for existence of root
        IF (FA * FB .LT. 0)THEN
! Add 1 to root count
            RCOUNT = RCOUNT + 1
! Exit if reached expected no of roots
            IF(RCOUNT .EQ. N)RETURN
! Update function value
            FA = FB
```

```
            ! Move to next root
            I = I + 1
            ! Assign current search pos. to new search pos.
            A(I) = B(I - 1)
         ELSE
            ! Update function value
            FA = FB
            ! Store current position
            A(I) = B(I)
         END IF
      END DO
   ! Arrive here if root count not met within defined interval
      PRINT *, 'Warning: Fewer roots found than predicted in interval'
      PRINT *, 'No of roots found:', RCOUNT
      M = RCOUNT
      RETURN
   END SUBROUTINE SEARCH
   !END OF FILE ***********************************************
```

finiteWell1.f90

```
!*************************************************************
MODULE UNITS
   DOUBLE PRECISION, PARAMETER :: H  = 6.62606957D0
   DOUBLE PRECISION, PARAMETER :: ME = 9.10938215D0
   DOUBLE PRECISION, PARAMETER :: EV = 1.602176565D0
   DOUBLE PRECISION, PARAMETER :: V0 = 1.D1, A = 5
   DOUBLE PRECISION  PI, HBAR2
```

END MODULE UNITS

!***

PROGRAM FINITEWELL1

  USE UNITS

  IMPLICIT NONE

  DOUBLE PRECISION, DIMENSION(3) :: AE, BE, AO, BO

  DOUBLE PRECISION, DIMENSION(6) :: E, ALPHA, BETA, C, D, PSI

  DOUBLE PRECISION, EXTERNAL   :: EVEN, ODD

  DOUBLE PRECISION, PARAMETER :: XMIN = -1.D1, XMAX = 1.D1

  DOUBLE PRECISION, PARAMETER :: XSTEP = 1.D-1

  DOUBLE PRECISION X, V

  INTEGER I

  OPEN(UNIT=1,FILE='rootfunc.txt')

  ! set the root brackets from plot of f(E)

  AE(1) = 1.D-1; BE(1) = 5.D-1

  AE(2) = 2.5D0; BE(2) = 3.D0

  AE(3) = 7.D0 ; BE(3) = 7.3D0

  AO(1) = 1.D0 ; BO(1) = 1.4D0

  AO(2) = 4.5D0; BO(2) = 4.8D0

  AO(3) = 9.6D0; BO(3) = 9.9D0

  PI = 4.D0*ATAN(1.D0)

  HBAR2 = H * H * 1.D2 / 4 / PI/ PI/ EV / ME

  ! We set ground state, which is even, in E(1) hence

  ! even I = odd function, odd I = even function

  DO I = 1, 3

    CALL BISECANT (AE(I), BE(I), E(2*I-1), EVEN)

    CALL BISECANT (AO(I), BO(I), E(2*I), ODD)

  END DO

  DO I = 1,6

Appendix

```
      ALPHA(I) = SQRT( 2 * E(I)/HBAR2 )


      BETA(I) = SQRT( 2 * (V0 - E(I))/HBAR2 )

      D(I) = SQRT( BETA(I)/(1+BETA(I)*A) )

      IF(MOD(I,2).EQ.0)THEN ! odd function

        C(I) = -D(I) * SIN(ALPHA(I)*A) * EXP(BETA(I)*A)

      ELSE ! even function

        C(I) = D(I) * COS(ALPHA(I)*A) * EXP(BETA(I)*A)

      END IF

    END DO

    X = XMIN

    DO WHILE ( X .LT. XMAX )

      DO I = 1, 6

        IF(X .LT. -A) THEN ! REGION I

          V = 1.D1

          PSI(I) = C(I)* EXP( BETA(I)*X )

        ELSE IF ( X .LE. A) THEN ! REGION II

          V = 0.D0

          IF(MOD(I,2).EQ.0) THEN ! odd function

            PSI(I) = D(I)*SIN(ALPHA(I)*X)

          ELSE ! even function

            PSI(I) = D(I)*COS(ALPHA(I)*X)

          END IF

        ELSE ! REGION III

          V = 1.D1

          IF(MOD(I,2).EQ.0)THEN !odd function

            PSI(I) = -C(I)*EXP( -BETA(I)*X )

          ELSE ! even function

            PSI(I) = C(I)*EXP( -BETA(I)*X )
```

```
      END IF
    END IF
  END DO
  WRITE(1,'(8F12.6)') X, V, (PSI(I)*PSI(I)+E(I), I=1,6)
  X = X + XSTEP
 END DO
 CLOSE(1)
END PROGRAM FINITEWELL1
```
!***
```
DOUBLE PRECISION FUNCTION EVEN(E)
 USE UNITS
 DOUBLE PRECISION ALPHA, BETA, E
 ! Mass = electron mass = 1
 ALPHA = SQRT( 2 * E/HBAR2 )
 BETA  = SQRT( 2 * (V0 - E)/HBAR2 )
 EVEN  = BETA * COS(ALPHA * A) - ALPHA * SIN(ALPHA * A)
END FUNCTION EVEN
```
!***
```
DOUBLE PRECISION FUNCTION ODD(E)
 USE UNITS
 DOUBLE PRECISION ALPHA, BETA, E
 ! Mass = electron mass = 1
 ALPHA = SQRT( 2 * E/HBAR2 )
 BETA  = SQRT( 2 * (V0 - E)/HBAR2 )
 ODD   = ALPHA * COS(ALPHA * A) + BETA * SIN(ALPHA * A)
END FUNCTION ODD
```
!**
```
SUBROUTINE BISECANT( A, B, R, F )
 DOUBLE PRECISION A, B, R, F
```

Appendix

```
DOUBLE PRECISION FA, FB, FR, DFR, DELTA, ERROR
INTEGER I
DOUBLE PRECISION, PARAMETER :: TOL = 1.D-08
! initialise variables
  FA = F(A)
  FB = F(B)
  ERROR = 1.D0
  I = 0
  IF(FA * FB .GT. 0)STOP 'Root not bracketed'
! Select a "best" starting point - arbitrary
  IF( ABS(FA) .LE. ABS(FB) ) THEN
     R = A
     FR = FA
  ELSE
     R = B
     FR = FB
  END IF
  DFR = (FB - FA)/(B - A)
! Set up the while loop to exit when the tolerance met
  DO WHILE( ERROR .GT. TOL )
! if iteration counter goes above 30 stop execution
     I = I + 1
     IF(I.GT.30)STOP 'Root not converged after 30 iters'
! Decide if using Secant or Bisection for this step
     IF( (DFR * (R - A) - FR)&
        *(DFR * (R - B) - FR).LE.0 )THEN ! Secant
        DELTA = -FR/DFR
        R = R + DELTA
     ELSE ! Bisection
```

```
        DELTA = (B - A)/2.D0

        R = (A + B)/2.D0

      END IF

! Evaluate function at the new value

      FR = F(R)

! Decide if root in left or right "half" of subinterval
! and adjust limits accordingly

      IF(FA * FR .LE. 0)THEN ! In left

        B = R

        FB = FR

      ELSE ! In right

        A = R

        FA = FR

      END IF

! Update the estimate of the derivative using the new value

      DFR = (FB - FA)/(B - A)

      ERROR = ABS(DELTA/R)

    END DO

    RETURN

  END SUBROUTINE

! END OF FILE **************************************
```

Appendix

Chapter 5 – Numerical Quadrature

adaptiveQuad.f90

```
!***************************************************************
! Program to perform an adaptive strip quadrature using the
! trapezoidal method.
! If T0 defines the numerical value of the trapezium rule
! integration with one strip, then T1 defines it with 2 strips
! (Tp, n = 2^p). Using the fact that the trapezium rule is of
! order h^2 we can obtain an estimate of the error using the
! formula (T0 - T1)/3. By comparing this estimate to the
! overall tolerance required we can either accept the numerical
! integration value, or halve the strip width and repeat
! the process until the desired accuracy is achieved.
!***************************************************************
  PROGRAM ADAPTIVEQUAD
    IMPLICIT NONE
  ! Declare variables
    DOUBLE PRECISION, PARAMETER :: EPS = 1.D-3
    DOUBLE PRECISION A, B, S, F
    EXTERNAL F
    OPEN(UNIT=1, FILE='adaptive_data.txt', ACTION='WRITE')
  ! Initialise integration limits
    A = 6.D0
    B = 1.4D1
  ! Initialise integration sum
    S = 0.D0
    CALL ADAPTIVE(F,A,B,S,EPS,0)
    PRINT *, 'Numerical integration =',S
```

```
      CLOSE(1)


   END PROGRAM ADAPTIVEQUAD
!****************************************************************
   DOUBLE PRECISION FUNCTION F(X)

      DOUBLE PRECISION X, I0, G, L0

         I0 = 1

         L0 = 10

         G  = 1

         F  = I0/(1 + 4*(X - L0)*(X - L0)/G/G)

      END FUNCTION F
!****************************************************************
   RECURSIVE SUBROUTINE ADAPTIVE(F,A,B,S,EPS,CT)

      DOUBLE PRECISION A,B,S,EPS

      DOUBLE PRECISION F,H,C,T0,T1

      INTEGER CT

      H = B - A  ! interval width for T0

      T0 = (F(B) + F(A)) * H/2

      H = H/2  ! halve the interval for T1

      C = (B + A)/2  ! midpoint

      T1 = H/2 * (2*F(C) + F(A) + F(B))

      CT = CT + 1

      IF (ABS( (T1-T0)/3 ).LE.EPS .AND. CT.GT.1 )THEN  ! Accept integration

         S = S + T1

         WRITE(1,100) A, F(A)! save left hand values for plotting
100      FORMAT(2(2X,F10.6))

      ELSE  ! Call subroutine with adjusted limits & eps

         CALL ADAPTIVE(F,A,C,S,EPS/2,CT)  ! left hand strip

         CALL ADAPTIVE(F,C,B,S,EPS/2,CT)  ! right hand strip
```

END IF

!Save upper limit and function evaluation for plotting

WRITE(1,100) B, F(B)

END SUBROUTINE ADAPTIVE

!END OF FILE***

Chapter6 – Ordinary Differential Equations
euler1.f90

PROGRAM EULER1A_DRIVER

  IMPLICIT NONE

  DOUBLE PRECISION, PARAMETER :: A = 0.D0

  DOUBLE PRECISION, PARAMETER :: B = 2.D0

  CHARACTER(LEN=13), PARAMETER :: FHEAD = 'euler1a_data_'

  CHARACTER(LEN=19) FNAME

  CHARACTER(LEN=2) NUM

  DOUBLE PRECISION X, Y

  DOUBLE PRECISION DF, SOL, ERR

  INTEGER N

  EXTERNAL DF, SOL

  DO N = 5,20,5

    WRITE(NUM,101) N

    FNAME = FHEAD//NUM//'.txt'

    Y = 1.D0

    CALL EULER1A(DF, X, Y, A, B, N, FNAME)

    ERR = ABS(SOL(X) - Y)

    PRINT*, 'H =',(B - A)/N, 'ERR =',ERR

  END DO

  101 FORMAT(I2.2)

END PROGRAM EULER1A_DRIVER

!***

DOUBLE PRECISION FUNCTION DF(X,Y)

  DOUBLE PRECISION X, Y

  DF = -X * Y

END FUNCTION

!***

DOUBLE PRECISION FUNCTION SOL(X)

  DOUBLE PRECISION X

  SOL = EXP(-X*X/2)

END FUNCTION

!***

SUBROUTINE EULER1A(DF, X, Y, A, B, N , FNAME)

  DOUBLE PRECISION X, Y, DF, H, A, B, SOL

  CHARACTER(*) FNAME

  INTEGER I, N

  IF(A.GT.B)STOP 'User error: B must be .gt. A'

  OPEN(UNIT=1, FILE=FNAME)

  X = A

  H = (B - A)/N

  WRITE(1,100) X, Y, SOL(X)

  DO I = 1,N-1

    Y = Y + H * DF(X,Y)

    X = X + H

    WRITE(1,100) X, Y, SOL(X)

```
      END DO
100 FORMAT(3(2X,F10.6))
      CLOSE(1)
      RETURN
      END SUBROUTINE EULER1A
!END OF FILE ********** *************************
```

rk4_mod.f90

```
      MODULE RK4_MODULE
      IMPLICIT NONE
      CONTAINS
!*******************************************************
      DOUBLE PRECISION FUNCTION DF(X,Y)
      DOUBLE PRECISION X, Y
      DF = -X*Y
      END FUNCTION DF
!*******************************************************
      DOUBLE PRECISION FUNCTION SOL(X)
      DOUBLE PRECISION X
      SOL = EXP(-X*X/2)
      END FUNCTION SOL
!*******************************************************
      SUBROUTINE RK4( DF, X, Y, A, B, N, FNAME )
      DOUBLE PRECISION X, Y, A, B
      DOUBLE PRECISION DF, H
      DOUBLE PRECISION K0, K1, K2, K3
      INTEGER I, N
      !Open file to save data
      CHARACTER(*) FNAME
```

```fortran
    OPEN(UNIT=1, FILE=FNAME)

    IF(A.GE.B)STOP 'User error: B must be .gt. A'

    !Initialise X and compute step size H

    X = A

    H = (B - A)/N

    !Write the initial values to file

    WRITE(1,100) X, Y, SOL(X)

    !Perform the RK4 integration

    DO I = 1, N-1

      !Compute Ks

      K0 = DF( X, Y )

      K1 = DF( X + H/2, Y + K0*H/2 )

      K2 = DF( X + H/2, Y + K1*H/2 )

      K3 = DF( X + H, Y + K2*H )

      !Advance one step

      Y = Y + (H/6)*(K0 + 2*K1 + 2*K2 + K3)

      X = X + H

      !Write each integration step values to file

      WRITE(1,100) X, Y, SOL(X)

    END DO

100 FORMAT(3(2X,F10.6))

    CLOSE(1)

    END SUBROUTINE RK4
!*****************************************************
    SUBROUTINE RK4A( DF, X, Y, A, B, TOL, FNAME )

    DOUBLE PRECISION X, Y, A, B, TOL

    DOUBLE PRECISION DF, H, Y0, Y1, X1, YT, XT

    DOUBLE PRECISION K0, K1, K2, K3, ERR

    INTEGER I
```

Appendix

```
CHARACTER(*) FNAME

OPEN(UNIT=1, FILE=FNAME)

IF(A.GE.B)STOP 'User error: B must be .gt. A'

!First step size equal to whole interval

X = A

H = B - A

!Save initial values to file

WRITE(1,101) X, Y, SOL(X)

!Set the condition to exit the while loop

DO WHILE ( (X.LT.B) )

  !Store values at start of step

  XT = X

  YT = Y

  !Compute Ks for y0

  K0 = DF( X, Y )

  K1 = DF( X + H/2, Y + K0*H/2 )

  K2 = DF( X + H/2, Y + K1*H/2 )

  K3 = DF( X + H, Y + K2*H )

  !Compute y0

  Y0 = Y + (H/6)*(K0 + 2*K1 + 2*K2 + K3)

  !Halve the step size

  H = H/2

  !Perform same integration using h/2

  DO I = 1,2

    K0 = DF( X, Y )

    K1 = DF( X + H/2, Y + K0*H/2 )

    K2 = DF( X + H/2, Y + K1*H/2 )

    K3 = DF( X + H, Y + K2*H )

    Y1 = Y + (H/6)*(K0 + 2*K1 + 2*K2 + K3)
```

```
            X1 = X + H

            Y = Y1

            X = X1

        END DO

        !Estimate the relative error in Y1

        ERR = ABS(Y1 - Y0)/Y1

        IF(ERR.GT.TOL)THEN

            !Go back to start of the step with h/2

            Y = YT

            X = XT

        ELSE

            !Accept the step using extrapolation

            Y = (16*Y - Y0)/15

            !Double the starting step size (currently h/2)

            H = H*4

            !X has already been advanced.

            WRITE(1,101) X, Y, SOL(X)

        END IF

    END DO

101 FORMAT(3(2X,F10.6))

    CLOSE(1)

    END SUBROUTINE RK4A

!**************************************************

END MODULE

!END OF FILE*************************************
```

euler2_mod.f90

```fortran
MODULE EULER2_MODULE

 IMPLICIT NONE

 CONTAINS

!****************************************************

   SUBROUTINE EULER2( DF, X, Y, A, B, N, FNAME)

   DOUBLE PRECISION DF, X, Y(2), A, B, K0(2), H

   INTEGER I, J, N

   EXTERNAL DF

   CHARACTER(*) FNAME

   IF(A.GE.B)STOP 'User error: B must be .gt. A'

   OPEN(UNIT=1, FILE=FNAME)

   X = A

   H = (B - A)/N

   !Write the initial values to file

   WRITE(1,100) X, Y(1), Y(2)

   DO I = 1,N-1

     CALL DERIV( X, Y, K0)

     DO J = 1,2

       Y(J) = Y(J) + H * K0(J)

     END DO

     X = X + H

     !Write each integration step values to file

     WRITE(1,100) X, Y(1), Y(2)

   END DO
100 FORMAT(3(2X,F10.6))

   CLOSE(1)

   END SUBROUTINE EULER2

!****************************************************
```

```fortran
SUBROUTINE DERIV( X, Y, K )

  DOUBLE PRECISION X, Y(2), K(2), DF

  EXTERNAL DF

  K(1) = Y(2)

  K(2) = DF( X, Y )

  RETURN

  END SUBROUTINE DERIV

END MODULE
!END OF FILE*****************************************
```

Chapter 7
dft_prog.f90

```fortran
!*****************************************************************
!Program to perform the discrete Fourier transform of the
!example function. Here we separate out the real and imaginary
!parts of the function and the transform. The inverse transform
!is also performed for comparison.
!*****************************************************************
PROGRAM DFT_EXAMPLE

  IMPLICIT NONE

  INTEGER, PARAMETER :: N=64,M=8

  DOUBLE PRECISION, DIMENSION (N) :: FR,FI,GR,GI

  DOUBLE PRECISION, EXTERNAL :: FUNC

  INTEGER I

  DOUBLE PRECISION F0,H,X

  CHARACTER(LEN=15), PARAMETER :: FNAME = 'dft_example.txt'

  OPEN(UNIT=1,FILE=FNAME)
```

Appendix

```fortran
! Initialise the variables.

! Here we compute a single factor of 1/N for convenience

Fo = 1.0/N

H  = 1.0/(N-1)

DO I = 1, N

   X = H*(I-1)

   FR(I) = FUNC(X) ! Re(fm)

   FI(I) = 0.0    ! Im(fm)

END DO

! Perform the Fourier transform

CALL DFT (FR,FI,GR,GI,N)

! Remember to include the factor 1/N

! We negate the imaginary part of g in preparation for

! the inverse transform.

DO I = 1, N

  GR(I) = Fo*GR(I)

  GI(I) = -Fo*GI(I)

END DO

! Perform the inverse transform

CALL DFT (GR,GI,FR,FI,N)

! The factor here is one as we have used 1/N previously

! Again negate the imaginary part to obtain the correct sign.

! Done for completeness.

DO I = 1, N

  FI(I) = -FI(I)

END DO

! Output the transformed data and function values

! to file in steps of M

WRITE (1,"(3F16.8)") (H*(I-1),FR(I),FUNC(H*(I-1))),I=1,N,M)
```

```fortran
! Capture the last data point
  WRITE (1,"(3F16.8)") H*(N-1),FR(N),FUNC(H*(N-1))
END PROGRAM DFT_EXAMPLE
!**************************************************
SUBROUTINE DFT (FR,FI,GR,GI,N)
! Subroutine to perform the discrete Fourier transform with
! FR and FI as the real and imaginary parts of the signal and
! GR and GI as the corresponding parts of the transform.
  IMPLICIT NONE
  INTEGER I,J,N
  DOUBLE PRECISION PI,X,Q
  DOUBLE PRECISION, DIMENSION (N) :: FR,FI
  DOUBLE PRECISION, DIMENSION (N) :: GR,GI
  PI = 4.D0*ATAN(1.D0)
  X = 2.D0*PI/N
  DO I = 1, N
    GR(I) = 0.0
    GI(I) = 0.0
    DO J = 1, N
      Q = X*(J-1)*(I-1)
      GR(I) = GR(I)+FR(J)*COS(Q)+FI(J)*SIN(Q)
      GI(I) = GI(I)+FI(J)*COS(Q)-FR(J)*SIN(Q)
    END DO
  END DO
END SUBROUTINE DFT
!**************************************************
DOUBLE PRECISION FUNCTION FUNC(X)
  DOUBLE PRECISION X, PI, SI, MU
! Gaussian distribution function
```

Appendix

```
    MU = 5.D-1 ! Mean

    SI = 1.D-1 ! sigma

    PI = 4.D0*ATAN(1.D0)

    FUNC = (1/SI/SQRT(2.D0*PI))*EXP(-(X-MU)*(X-MU)/2.D0/SI/SI)
END FUNCTION FUNC
!END OF FILE*****************************************************
```

fft_mod.f90

```
MODULE FFT_MOD
IMPLICIT NONE
CONTAINS
!****************************************************************
SUBROUTINE FFT (A,M,INV)
    ! An example of the fast Fourier transform subroutine with N = 2**M.
    ! A(N) is the complex data in the input and corresponding Fourier
    ! coefficients in the output. INV is flag to compute inverse(1)
    ! transform or not(0).
    IMPLICIT NONE
    INTEGER N,N2,M,I,J,K,L,L1,L2,IP,INV
    DOUBLE PRECISION PI, Q, WR, WI
    COMPLEX*16 :: A(*), U, W, T
    PI = 4.D0*ATAN(1.D0)
    N = 2**M
    N2 = N/2
    ! Rearrange the data to the bit reversed order
    ! This interweaves the even and odd function evaluations
    J = 1
    DO I = 1, N-1
        IF (I.LT.J) THEN
```

```
      T    = A(J)
      A(J) = A(I)
      A(I) = T
    END IF
    K  = N2
    DO WHILE (K.LT.J)
      J = J - K
      K = K/2
    END DO
    J = J + K
  END DO
  ! Perform additions at all levels with reordered data
  ! i.e. perform the FFT
  L2 = 1
  DO L = 1, M
    L1 = L2
    L2 = 2*L2
    Q = PI/L1
    U = ( 1.D0, 0.D0 )
    WR = COS(Q)
    WI = SIN(Q)
    W = DCMPLX( WR, -WI )
    IF(INV .EQ. 1) W = DCONJG(W) ! complex conjugate
    DO J = 1, L1
      DO I = J, N, L2
        IP   = I + L1
        T    = A(IP)*U
        A(IP) = A(I)-T
        A(I) = A(I)+T
```

Appendix

```
      END DO


    U = U * W

  END DO

END DO

! Apply the factor of 1/N if not taking the inverse transform

IF(INV .NE. 1) THEN

  DO I = 1, N

    A(I) = A(I)/N

  END DO

END IF

END SUBROUTINE FFT

END MODULE FFT_MOD

!END OF FILE **************************************************
```

Chapter 8 – Monte Carlo Methods

PI_Monte.f90

```
PROGRAM PI_MONTE

IMPLICIT NONE

!------------------------------------------------------------------
! A program to calculate pi using the dart theory. A dart is randomly
! thrown into the unit square, x [0,1], y [0,1].
! From the comparison of the areas of the square and
! the quarter circle, the chance of the dart landing within a unit circle
! centred on the origin is pi/4.
!------------------------------------------------------------------
!------------------------------------------------------------------

INTEGER,    PARAMETER :: N = 1000, C = 10
```

```fortran
      DOUBLE PRECISION, PARAMETER :: ONE = 1.0D+0
      INTEGER        :: I, M
      DOUBLE PRECISION :: R(2), NORM, PIE, PI
      ! n - total number of darts to throw
      ! c - print out pie every c darts thrown
      ! i - iteration counter; number of darts thrown.
      ! m - number of darts successfully landed within quarter circle.
      ! r - position vector representing the coordinates of where the dart lands
      !     in the xy plane
      ! norm - size of the position vector from the origin. norm.le.1.0 dart lands
      !       within quarter circle, else lands outside.
      ! pie - approximation of pi.
      ! pi - actual pi
      OPEN(UNIT=1, FILE='PIdata.txt') ! file for pi approximation
      OPEN(UNIT=2, FILE='RanXY.txt') ! file to check "randomness" of throws
      PI = 4.D0*ATAN(1.D0)
10    FORMAT(2F12.6)
11    FORMAT(I10, 5X, 2F11.8)
      ! Initialise variables
      M = 0
      CALL INIT_RANDOM_SEED()
      DO I = 1,N
        ! Generate random position vector
        CALL RANDOM_NUMBER(R) ! populates R with 2 random numbers.
        IF(MOD(I,C).EQ.0) THEN
           WRITE (2,10) R(1), R(2) ! print random numbers to file
        END IF
        NORM = R(1)*R(1) + R(2)*R(2) ! compute distance from origin
        IF(NORM.LE.ONE) THEN ! hit the circle add one to counter
```

Appendix

```fortran
      M = M + 1
   END IF
   IF(MOD(I,C).EQ.0) THEN ! Print out every C dart thrown.
      PIE = 4.0D0*M/I ! ensure double precision
      WRITE (1,11) I, PIE, PI
   END IF
END DO
CLOSE(1)
CLOSE(2)
END PROGRAM PI_MONTE
!**************************************************************
SUBROUTINE INIT_RANDOM_SEED()
   INTEGER :: I, N
   REAL    :: T
   INTEGER, DIMENSION(:), ALLOCATABLE :: SEED
   CALL RANDOM_SEED(SIZE = N)
   ALLOCATE( SEED(N) )
   T = SECNDS(0.0)
   SEED = INT(T) + 37 * (/ (I - 1, I = 1, N) /)
   DO I =1,N
      IF (MOD(SEED(I),2).EQ.0) SEED(I) = SEED(I) -1
   END DO
   CALL RANDOM_SEED(PUT = SEED)
   DEALLOCATE(SEED)
END SUBROUTINE INIT_RANDOM_SEED
!END OF FILE **************************************************
```

monte_carlo.f90

```fortran
MODULE MOD_MC

  IMPLICIT NONE

  CONTAINS
!*****************************************************************************

  SUBROUTINE MONTE_CARLO(A, B, N, F,SUM,SIG)

    IMPLICIT NONE

    INTEGER        :: I, N

    DOUBLE PRECISION :: F, A, B, R, SUM, SUM2, SIG, S, S2

    EXTERNAL F

    IF(N.LE.0)STOP "User error: N must be .gt. zero"

    IF(A.GT.B)STOP "User error: A must be .lt. B"

    SUM = 0.D0

    SUM2 = 0.D0

    SIG = 0.D0

    CALL INIT_RANDOM_SEED()

    DO I = 1,N

      CALL RANDOM_NUMBER(R)

      R = A + (B - A)* R

      SUM = SUM + F(R)

      SUM2 = SUM2 + F(R)*F(R)

      S = SUM/DBLE(I)

      S2 = SUM2/DBLE(I)

      IF(I.NE.1)SIG = SQRT( (S2 - (S*S))/(DBLE(I)-1.D0) )

    END DO

    SUM = (B - A)*SUM/N

    SIG = (B - A)*SIG

  END SUBROUTINE MONTE_CARLO
```

Appendix

!***

```
SUBROUTINE INIT_RANDOM_SEED()

  INTEGER :: I, N

  REAL   :: T

  INTEGER, DIMENSION(:), ALLOCATABLE :: SEED

  CALL RANDOM_SEED(SIZE = N)

  ALLOCATE( SEED(N) )

  T = SECNDS(0.0)

  SEED = INT(T) + 37 * (/ (I - 1, I = 1, N) /)

  DO I =1,N

    IF (MOD(SEED(I),2).EQ.0) SEED(I) = SEED(I) -1

  END DO

  CALL RANDOM_SEED(PUT = SEED)

  DEALLOCATE(SEED)

 END SUBROUTINE INIT_RANDOM_SEED
END MODULE MOD_MC
```

!END OF FILE ***

monte_decay.f90

```
PROGRAM MONTE_DECAY

 IMPLICIT NONE

 !------------------------------------------------------------

 INTEGER,    PARAMETER :: MAX = 1000, T_MAX = 500

 DOUBLE PRECISION, PARAMETER :: LAMBDA = 0.01

 INTEGER      :: ATOM, T, N, NCOUNT

 DOUBLE PRECISION  :: DECAY

 OPEN(UNIT=1, FILE='decay_data.txt')
10 FORMAT(2(I5))

 ! Initialise variables
```

```
N = MAX

NCOUNT = MAX

CALL INIT_RANDOM_SEED()

DO T = 1,T_MAX

  DO ATOM = 1,N

    CALL RANDOM_NUMBER(DECAY)

    IF(DECAY.LT.LAMBDA) NCOUNT = NCOUNT - 1

  END DO

  N = NCOUNT

  WRITE(1,10) T, N

END DO

CLOSE(1)

STOP

END PROGRAM MONTE_DECAY
```

!***
!END OF FILE ***

Chapter 9 – Partial Differential Equations

gauss-seidel.f90

```
SUBROUTINE GAUSS_SIEDEL( F, N, A, B, C, D, X1, XN, TOL, FNAME)

  IMPLICIT NONE

  ! F(N) on entry contains initial guess, on exit contains iterated value

  ! N dimension of the problem; H = (XN - X1)/(N-1)

  ! A thru D are the coefficients of the general second order ODE

  !      aF" + bF' + cF = dX

  ! X1, XN start and end of computational domain on x

  ! TOL tolerance we want in the iteration will exit once met or itcount > itmax
```

Appendix

```
! FNAME file name to which to store data
INTEGER :: N
DOUBLE PRECISION :: F(N), A, B, C, D, X1, XN, TOL
CHARACTER(*) :: FNAME
LOGICAL DONE
DOUBLE PRECISION H, FF, THE, PSI, PHI, X
INTEGER, PARAMETER :: ITMAX = 100
INTEGER I, J, ITCOUNT
DOUBLE PRECISION Z(ITMAX+1,N)
OPEN(UNIT=1,FILE=FNAME)
100 FORMAT(9999(2X,F10.6))
! calculate parameters and initialise variables
H = (XN - X1)/(N-1)
THE = C - 2*A/H/H
PHI = (A/H/H) + (B/2/H)
PSI = (A/H/H) - (B/2/H)
ITCOUNT = 0
DONE = .FALSE.
! perform the iteration loop
DO WHILE (DONE .EQV. .FALSE. .AND. ITCOUNT .LE. ITMAX)
    DONE = .TRUE.
    ITCOUNT = ITCOUNT + 1
    DO I = 2, N -1
        X = X1 + H * (I - 1)
        FF = -( PHI * F(I+1) + PSI * F(I-1) -D*X )/THE
        IF ( ABS( (FF - F(I))/FF ) .GT. TOL ) DONE = .FALSE.
        F(I) = FF
        ! use z to print data to file
        Z(ITCOUNT, I) = F(I)
```

```fortran
      END DO

   END DO

! print message to screen if tolerance not reached

   IF(ITCOUNT .GT. ITMAX) THEN

      PRINT *, 'Iteration count gone above iteration maximum'

   END IF

! add x1 an xn values to z

   DO I = 1,ITCOUNT

      Z(I, 1) = F(1)

      Z(I, N) = F(N)

   END DO

! print results to file

   DO I = 1,ITCOUNT

      WRITE(1,100) (Z(I,J), J=1,N)

   END DO

! print to screen iteration count

   PRINT *, 'Number of iterations =',ITCOUNT

   CLOSE(1)

END SUBROUTINE GAUSS_SIEDEL

!END OF FILE*************************************************
```

richardsonExtrap.f90

```fortran
PROGRAM RICHARDSON_EXTRAP

   IMPLICIT NONE

   DOUBLE PRECISION, PARAMETER :: TOL = 1.D-4, X = 1.D0

   INTEGER,         PARAMETER :: ROW_MAX = 4

   DOUBLE PRECISION FUNC, Y, H, A(ROW_MAX, ROW_MAX)

   INTEGER M, L, I, J

   LOGICAL SOLUTION
```

Appendix

```
EXTERNAL FUNC

H = 4.D-1

A = 0.D0

SOLUTION = .FALSE.

CALL CENTRALDIFF(X, Y, H, FUNC)

A(1,1) = Y

DO M = 1, ROW_MAX - 1

   H = H/2

   CALL CENTRALDIFF(X, Y, H, FUNC)

   A(M + 1, 1) = Y

   DO L = 1,M

      A(M + 1, L + 1) = (2**(2*L) * A(M + 1, L) - A(M, L))/(2**(2*L) - 1)

   END DO

   IF(ABS(A( M+1, M+1 ) - A( M, M )) .LT. TOL) THEN

      PRINT *, 'Solution found within tolerance'

      DO I = 1,M+1

         WRITE (*,"(10F14.10)") ( A(I,J), J = 1, M + 1)

         SOLUTION = .TRUE.

      END DO

      EXIT

   END IF

END DO

IF(SOLUTION .EQV. .FALSE.) THEN

   PRINT *, "Extrapolation not found within tolerance specified"

   DO I = 1,M+1

      WRITE (*,"(10F14.10)") ( A(I,J), J = 1, M + 1)

   END DO

END IF

END PROGRAM RICHARDSON_EXTRAP
```

```fortran
!*******************************************************************
SUBROUTINE CENTRALDIFF (X, Y, H, FUNC)
  DOUBLE PRECISION X, Y, H, FUNC
  EXTERNAL FUNC
  Y = (FUNC( X + H ) - FUNC ( X - H ))/2/H
END SUBROUTINE CENTRALDIFF
!*******************************************************************
DOUBLE PRECISION FUNCTION FUNC(X)
  DOUBLE PRECISION X
  FUNC = SIN(X)
END FUNCTION FUNC
!END OF FILE ******************************************************
```

heatEqn1.f90

```fortran
PROGRAM HEAT_EQUATION
  IMPLICIT NONE
  INTEGER, PARAMETER :: M = 6, N = 101
  DOUBLE PRECISION :: AD(M-2), AS(M-3), B(M-2,M), U(M,N), US(M-2)
  DOUBLE PRECISION :: AD2(M-2), AS2(M-3)
  DOUBLE PRECISION H, K, R, W, L, T_END, D1, D2
  INTEGER I, J, INFO
  W = 5.D-1 ! Crank-Nicolson
  L = 1.D0 ! length of conducting rod
  D1 = 1.D2 ! Dirichlet boundary condtion at x = 0
  D2 = 1.D2 ! Dirichlet boundary condtion at x = L
  T_END = 1.D0 ! time to run system
  H = L/(M-1) ! H = 0.2
  K = T_END/(N-1) ! K = 0.01
```

Appendix

R = K/H/H ! R = 0.01/0.2**2 = 1/4

! Initialise our variables ****************************

!Set up the diagonal of A.

!Store to another array for later use.

DO I = 1, M-2

   AD(I) = 1 + 2 * R * W

   AD2(I) = AD(I)

END DO

!set up the sub/super diagonal of A

!Store to another array for later use.

DO I = 1, M-3

   AS(I) = -R * W

   AS2(I) = AS(I)

END DO

!Initialise B to zero

B = 0.D0

!Set up non-zero values of B

DO I = 1, M -2

   B(I,I)   = R * (1 - W)

   B(I,I + 1) = 1 - 2 * R *(1 - W)

   B(I,I + 2) = R * (1 - W)

END DO

!Adjust first and last entries for boundary values

B(1,1) = R

B(M-2,M) = R

!Initial conditions: U(x,0) = 0.d0 Celsius

! set the entire system to the IC, advanced time step

! values will be overwritten.

U = 0.D0

```fortran
!Account for boundary conditions
DO J = 1,N
   U(1,J) = D1
   U(M,J) = D2
END DO
! Main loop ******************************************
DO J = 1, N-1
   !Compute the known right hand sides
   DO I = 1, M-2
      US(I) = B(I,I)*U(I,J) + B(I,I+1)*U(I+1,J) + &
         B(I,I+2)*U(I+2,J)
   END DO
   !Solve the tridiagonal system using LAPACK routine DPTSV
   CALL DPTSV(M-2, 1, AD, AS, US, M-2, INFO)
   !Write the solution back to U at the advanced time j+1
   !Remember to write to interior grid points i=2, m-1.
   DO I = 1, M-2
      U(I+1,J+1) = US(I)
   END DO
   !As the Lapack subroutine overwrites the arrays AD and AS
   !with their factors we need to reset these.
   !NOTE: THIS IS NOT OPTIMAL. How can it be improved? Hint:
   !Take a look at the DPTSV source code located at
   !http://www.netlib.org/lapack/double/dptsv.f
   DO I = 1, M-2
      AD(I) = AD2(I)
   END DO
   DO I = 1, M-3
      AS(I) = AS2(I)
```

```
        END DO

    END DO

! Extract data to file *******************************

    OPEN(UNIT=1, FILE='heat_eqn_data.txt')

    DO I = 1,M

        WRITE (1,'(1001f10.4)') ( U(I,J), J=1,N )

    END DO

    CLOSE(1)

END PROGRAM HEAT_EQUATION

!END OF FILE *****************************************
```

Chapter 10 – Advanced Numerical Quadrature

GaussLeg.f90

```
! Program to integrate f(x) = exp(x) on the interval [-1,4]
! using Gauss-Legendre quadrature.
! Here we set quadruple precision
PROGRAM GAUSS

    IMPLICIT NONE

    INTEGER, PARAMETER :: p = 16 ! quadruple precision

    INTEGER         :: N = 10, K

    REAL(kind=p), ALLOCATABLE :: R(:,:)

    REAL(kind=p)    :: Z, A, B, EXACT

    ! Perform N point quadrature

    DO N = 1,20

        A = 1.0_p

        B = 4.0_p

        ! R contains abscissas in 1st row and weights in 2nd row
```

```
R = GAUSSQUAD(N)
! Compute the integral the dot_product(j,k) = sum over n (jk)
Z = (B-A)/2._p*DOT_PRODUCT(R(2,:),EXP((B+A)/2._p+R(1,:)*(B-A)/2._p))
! Calculate the exact value
EXACT = EXP(B)-EXP(A)
! Print to screen the results for each quadrature.
PRINT "(I0,1X,G0,1X,G10.2)",N, Z, Z-EXACT
END DO

CONTAINS
! Function computes Legendre polynomials, their roots
! and corresponding weights
FUNCTION GAUSSQUAD(N) RESULT(R)
    INTEGER         :: N
    REAL(kind=p), PARAMETER :: PI = 4._p*ATAN(1._p)
    REAL(kind=p)    :: R(2, N), X, F, DF, DX
    INTEGER         :: I, ITER
    REAL(kind = p), ALLOCATABLE :: P0(:), P1(:), TMP(:)
    ! P0, P1, and TMP contain the polynomial coefficients
    ! they expand with K up to N. The [] concatenates the arrays.
    P0 = [1._p]
    P1 = [1._p, 0._p]
    ! compute polynomial coefficients
    DO K = 2, N
        TMP = ((2*K-1)*[P1,0._p]-(K-1)*[0._p, 0._p,P0])/K
        P0 = P1
        P1 = TMP
    END DO
    DO I = 1, N
        ! Compute initial guess
```

Appendix

```
        X = COS(PI*(I-0.25_p)/(N+0.5_p))

        DX = 1._p

        ITER = 1

        ! perform Newton Raphson root search

        ! here we compute F and DF for the given X

        ! EXIT once root reaches sufficient precision

        ! OR the iteration count gt 10.

        DO WHILE (ABS(DX).GT.10.0*EPSILON(DX) .AND. ITER .LE. 10)

          F = P1(1)

          DF = 0._p

          !as P1 can expand we set the upper limit as the
          !size of P1.

          DO K = 2, SIZE(P1)

            DF = F + X*DF

            F  = P1(K) + X * F

          END DO

          DX = F / DF

          X = X - DX

          ITER = ITER + 1

        END DO

        !once root is found save to array R and compute weight

        R(1,I) = X

        R(2,I) = 2.0_p/((1.0_p-X*X)*DF*DF)

      END DO

    END FUNCTION GAUSSQUAD

  END PROGRAM GAUSS

!END OF FILE *************************************************
```

Chapter 11 – An Advanced ODE Solver
rkf45N_mod.f90

```fortran
!*************************************************************
!Module containing Runge-Kutta-Fehlberg algorithms to deal
!with both first (RK4F1) and second ordered (RK4FN) ODEs.
!The 2nd ordered ODEs are reduced to coupled 1st order ODEs.
!The R-K methods used are fourth/fifth order.
!The subroutine DERIV calculates the derivatives of all
!components of the derivative vector
!*************************************************************
MODULE RKF45_MODULE
  IMPLICIT NONE
  CONTAINS
   !*******************************************************
   SUBROUTINE RKF451( DF, X, Y, A, B, TOL )
   !*******************************************************
   !RK4F1 uses a fourth/fifth order Runge-Kutta-Fehlberg
   !algorithm to solve first order ODEs.
   !*******************************************************
   !Declare subroutine arguments
   DOUBLE PRECISION DF, X, Y, A, B, TOL
   EXTERNAL DF
   !Adjustable parameters *****************************
   !Define the conservative factor for new step size
   DOUBLE PRECISION, PARAMETER :: ALPHA = 9.D-1
   !FRACMAX: Fraction of interval defining HMAX (B-A/FRACMAX)
   !FRACMIN: Fraction of interval defining HMIN (B-A/FRACMIN)
   DOUBLE PRECISION, PARAMETER :: FRACMAX = 1.D1
```

Appendix

DOUBLE PRECISION, PARAMETER :: FRACMIN = 1.D5

!INC: Max tolerated increase factor of h

!DEC: Max tolerated decrease factor of h

DOUBLE PRECISION, PARAMETER :: INC = 4.D0

DOUBLE PRECISION, PARAMETER :: DEC = 1.D-1

!DO NOT CHANGE the coefficients below ***************

! Am define the coefficients used to calculate x

DOUBLE PRECISION, PARAMETER :: A1 = 2.5D-1

DOUBLE PRECISION, PARAMETER :: A2 = 3.75D-1

DOUBLE PRECISION, PARAMETER :: A3 = 1.2D1/1.3D1

DOUBLE PRECISION, PARAMETER :: A4 = 1.D0

DOUBLE PRECISION, PARAMETER :: A5 = 5.D-1

!Bmn define the coefficients used to calculate y

DOUBLE PRECISION, PARAMETER :: B10 = 2.5D-1

DOUBLE PRECISION, PARAMETER :: B20 = 3.D0/3.2D1

DOUBLE PRECISION, PARAMETER :: B21 = 9.D0/3.2D1

DOUBLE PRECISION, PARAMETER :: B30 = 1.932D3/2.197D3

DOUBLE PRECISION, PARAMETER :: B31 = -7.2D3/2.197D3

DOUBLE PRECISION, PARAMETER :: B32 = 7.296D3/2.2197D3

DOUBLE PRECISION, PARAMETER :: B40 = 4.39D2/2.16D2

DOUBLE PRECISION, PARAMETER :: B41 = -8.D0

DOUBLE PRECISION, PARAMETER :: B42 = 3.68D3/5.13D2

DOUBLE PRECISION, PARAMETER :: B43 = -8.45D2/4.104D3

DOUBLE PRECISION, PARAMETER :: B50 = -8.D0/2.7D1

DOUBLE PRECISION, PARAMETER :: B51 = 2.D0

DOUBLE PRECISION, PARAMETER :: B52 = -3.544D3/2.565D3

DOUBLE PRECISION, PARAMETER :: B53 = 1.859D3/4.104D3

DOUBLE PRECISION, PARAMETER :: B54 = -1.1D1/4.D1

!Cm are used to evaluate YHAT

```fortran
DOUBLE PRECISION, PARAMETER :: C0 = 1.6D1/1.35D2
DOUBLE PRECISION, PARAMETER :: C2 = 6.656D3/1.2825D4
DOUBLE PRECISION, PARAMETER :: C3 = 2.8561D4/5.643D4
DOUBLE PRECISION, PARAMETER :: C4 = -9.D0/5.D1
DOUBLE PRECISION, PARAMETER :: C5 = 2.D0/5.5D1
!Dm are used to calculate the error
DOUBLE PRECISION, PARAMETER :: D0 = 1.D0/3.6D2
DOUBLE PRECISION, PARAMETER :: D2 = -1.28D2/4.275D3
DOUBLE PRECISION, PARAMETER :: D3 = -2.197D3/7.524D4
DOUBLE PRECISION, PARAMETER :: D4 = 1.D0/5.D1
DOUBLE PRECISION, PARAMETER :: D5 = 2.D0/5.5D1
!Declare local variables
DOUBLE PRECISION H, HMAX, HNEW, HMIN, X0, Y0, YHAT
DOUBLE PRECISION K0, K1, K2, K3, K4, K5, ERR
IF(A.GE.B)STOP 'User error: B must be .gt. A'
IF(TOL.LT.EPSILON(0.D0))THEN
   PRINT*, 'User error: Tol must be .gt.',EPSILON(0.D0)
   STOP
END IF
!Open a file to save data
OPEN(UNIT=1, FILE='rk4f_data.txt')
!Set maximum step size 1/10 of full range
HMAX = (B - A)/FRACMAX
!Set minimum step size 1/10,000 of full range
HMIN = (B - A)/FRACMIN
!Initialise the integration;
!initial y defined in calling program.
H = HMAX
X0 = A
```

Appendix

```
Y0 = Y

!Perform the integration

DO WHILE (X0.LE.B)

  K0 = DF(X0,Y0)

  X = X0 + A1*H

  Y = Y0 + B10*H*K0

  K1 = DF(X,Y)

  X = X0 + A2*H

  Y = Y0 + H*(B20*K0 + B21*K1)

  K2 = DF(X,Y)

  X = X0 + A3*H

  Y = Y0 + H*(B30*K0 + B31*K1 + B32*K2)

  K3 = DF(X,Y)

  X = X0 + A4*H

  Y = Y0 + H*(B40*K0 + B41*K1 + B42*K2 + B43*K3)

  K4 = DF(X,Y)

  X = X0 + A5*H

  Y = Y0 + H*(B50*K0 + B51*K1 + B52*K2 + B53*K3 &
    + B54*K4)

  K5 = DF(X,Y)

  YHAT = Y0 + H*(C0*K0 + C2*K2 + C3*K3 + C4*K4 &
    + C5*K5)

  ERR = H * ABS(D0*K0 + D2*K2 + D3*K3 + D4*K4 &
    + D5*K5)

  HNEW = ALPHA * H * SQRT( SQRT( H*TOL/ERR ) )

  !Check prediction of step size falls within
  !acceptable limits and adjust if necessary

  IF( HNEW .GT. INC * H )HNEW = INC * H

  IF( HNEW .LT. DEC * H )HNEW = DEC * H
```

```
      IF( HNEW .GT. HMAX)HNEW = HMAX
      IF( HNEW .LT. HMIN)THEN
        PRINT *, H,'->', HNEW
        PRINT *,'Step size too small; exiting integration'
        WRITE (1,100) X, YHAT
        STOP 'Possible problem with RK4F subroutine'
      END IF
      !Check error vs. max_error = h*eps
      !if error too large repeat step using hnew
      !else accept propagation
      IF( ERR .GT. (H*TOL) )THEN
        H = HNEW
      ELSE
        Xo = Xo + H
        Yo = YHAT
        H = HNEW
        !Save the data to file
        WRITE(1,100) Xo, Yo
      END IF
    END DO
100 FORMAT(2(2X,F10.6))
    !Assign calculated values back to subroutine arguments
    X = Xo
    Y = Yo
    CLOSE(1)
    RETURN
    END SUBROUTINE RKF451

!*************************************************************

    SUBROUTINE RKF45N( N, DF, X, Y, A, B, TOL, FNAME, FREQ )
```

Appendix

```
!*********************************************************
!RK4FN uses a fourth/fifth order Runge-Kutta-Fehlberg
!algorithm to solve second order ODEs.
!2nd order ODE -> pair of coupled 1st order ODEs
!for any number of dependent functions
!backwards integration is allowed i.e. B .lt. A
!*********************************************************
!Declare subroutine arguments
INTEGER FREQ, N
DOUBLE PRECISION DF, X, Y(N), A, B, TOL
CHARACTER(*) FNAME
EXTERNAL DF
!Define the conservative factor for new step size
DOUBLE PRECISION, PARAMETER :: ALPHA = 9.D-1
! Am define the coefficients used to calculate x
DOUBLE PRECISION, PARAMETER :: A1 = 2.5D-1
DOUBLE PRECISION, PARAMETER :: A2 = 3.75D-1
DOUBLE PRECISION, PARAMETER :: A3 = 1.2D1/1.3D1
DOUBLE PRECISION, PARAMETER :: A4 = 1.D0
DOUBLE PRECISION, PARAMETER :: A5 = 5.D-1
!Bmn define the coefficients used to calculate y
DOUBLE PRECISION, PARAMETER :: B10 = 2.5D-1
DOUBLE PRECISION, PARAMETER :: B20 = 3.D0/3.2D1
DOUBLE PRECISION, PARAMETER :: B21 = 9.D0/3.2D1
DOUBLE PRECISION, PARAMETER :: B30 = 1.932D3/2.197D3
DOUBLE PRECISION, PARAMETER :: B31 = -7.2D3/2.197D3
DOUBLE PRECISION, PARAMETER :: B32 = 7.296D3/2.2197D3
DOUBLE PRECISION, PARAMETER :: B40 = 4.39D2/2.16D2
DOUBLE PRECISION, PARAMETER :: B41 = -8.D0
```

```fortran
DOUBLE PRECISION, PARAMETER :: B42 = 3.68D3/5.13D2

DOUBLE PRECISION, PARAMETER :: B43 = -8.45D2/4.104D3

DOUBLE PRECISION, PARAMETER :: B50 = -8.D0/2.7D1

DOUBLE PRECISION, PARAMETER :: B51 = 2.D0

DOUBLE PRECISION, PARAMETER :: B52 = -3.544D3/2.565D3

DOUBLE PRECISION, PARAMETER :: B53 = 1.859D3/4.104D3

DOUBLE PRECISION, PARAMETER :: B54 = -1.1D1/4.D1

!Cm are used to evaluate YHAT

DOUBLE PRECISION, PARAMETER :: C0 = 1.6D1/1.35D2

DOUBLE PRECISION, PARAMETER :: C2 = 6.656D3/1.2825D4

DOUBLE PRECISION, PARAMETER :: C3 = 2.8561D4/5.643D4

DOUBLE PRECISION, PARAMETER :: C4 = -9.D0/5.D1

DOUBLE PRECISION, PARAMETER :: C5 = 2.D0/5.5D1

!Dm are used to calculate the error

DOUBLE PRECISION, PARAMETER :: D0 = 1.D0/3.6D2

DOUBLE PRECISION, PARAMETER :: D2 = -1.28D2/4.275D3

DOUBLE PRECISION, PARAMETER :: D3 = -2.197D3/7.524D4

DOUBLE PRECISION, PARAMETER :: D4 = 1.D0/5.D1

DOUBLE PRECISION, PARAMETER :: D5 = 2.D0/5.5D1

!Declare local variables

DOUBLE PRECISION H, HMAX, HNEW, HMIN, X0, ERR, BIGERR

DOUBLE PRECISION, DIMENSION(N) :: Y0, YHAT, K0, K1

DOUBLE PRECISION, DIMENSION(N) :: K2, K3, K4, K5

INTEGER I, COUNT

IF(MOD(N,2).NE.0) STOP 'User error: N must be even'

IF(TOL.LT.EPSILON(0.D0)) THEN

   PRINT*, 'User error: Tol must be .gt.',EPSILON(0.D0)

   STOP

END IF
```

Appendix

```
IF(FREQ.GT.1000)THEN
    STOP 'User error: only a MAX of 1000 data points allowed'
END IF

!Open a file to save data
OPEN(UNIT=1, FILE=FNAME)

!Initialise print count
COUNT = 1

!Set maximum step size as 1/freq of full range
!i.e. freq is the number of points to save
HMAX = (B - A)/FREQ

!Set minimum step size to a reasonable (+ve) value say,
HMIN = 1.D1*EPSILON(0.D0)

!Initialise the integration;
!initial Y values defined in calling program.
H = HMAX

X0 = A

DO I = 1,N
    Y0(I) = Y(I)
END DO

!Save the initial conditions
WRITE(1,100) X0, (Y0(I), I=1,N)

!Perform the integration
DO WHILE ( ABS(X0-A).LT.ABS(B-A) )
    CALL DERIV( N, X0, Y0, K0 )
    X = X0 + A1*H
    DO I = 1,N
        Y(I) = Y0(I) + B10*H*K0(I)
    END DO
    CALL DERIV( N, X, Y, K1 )
```

```
X = X0 + A2*H
DO I = 1,N
   Y(I) = Y0(I) + H*(B20*K0(I) + B21*K1(I))
END DO
CALL DERIV( N, X, Y, K2 )
X = X0 + A3*H
DO I = 1,N
   Y(I) = Y0(I) + H*(B30*K0(I) + B31*K1(I) &
     + B32*K2(I))
END DO
CALL DERIV( N, X, Y, K3 )
X = X0 + A4*H
DO I = 1,N
   Y(I) = Y0(I) + H*(B40*K0(I) + B41*K1(I) &
     + B42*K2(I) + B43*K3(I))
END DO
CALL DERIV( N, X, Y, K4 )
X = X0 + A5*H
DO I = 1,N
   Y(I) = Y0(I) + H*(B50*K0(I) + B51*K1(I) &
     + B52*K2(I) + B53*K3(I) + B54*K4(I))
END DO
CALL DERIV( N, X, Y, K5 )
BIGERR = 0.D0
DO I = 1,N
   YHAT(I) = Y0(I) + H*(C0*K0(I) + C2*K2(I) &
     + C3*K3(I) + C4*K4(I) + C5*K5(I))
   ERR = H * ABS( D0*K0(I) + D2*K2(I) + D3*K3(I) &
     + D4*K4(I) + D5*K5(I) )
```

Appendix

```
      IF( ABS(ERR) .GT. ABS(BIGERR) )BIGERR = ERR
   END DO
   HNEW = ALPHA * H * SQRT( SQRT( H*TOL/BIGERR ) )


   !Check prediction of step size falls within
   !acceptable limits and adjust if necessary
   IF( ABS(HNEW) .GT. 4.D0 * ABS(H) ) HNEW = 4.D0 * H
   IF( ABS(HNEW) .LT. .1D0 * ABS(H) ) HNEW = .1D0 * H
   IF( ABS(HNEW) .GT. ABS(HMAX) )    HNEW = HMAX
   ! No ABS required for HMIN as it is a +ve value
   IF( ABS(HNEW) .LT. HMIN)THEN
      PRINT *, H,'->', HNEW
      WRITE (1,100) X, (YHAT(I),I=1,N)
      STOP 'Step size too small; exiting integration'
   END IF
   !Check error vs. Max error = h*eps
   !if error too large repeat step using hnew
   !else accept propagation
   IF( ABS(BIGERR) .GT. ABS(H)*TOL )THEN
      !check new step doesn't take us past B
      IF(ABS(X0 - A + HNEW).GT. ABS(B - A)) HNEW = B - X0
      H = HNEW
   ELSE
      X0 = X0 + H
      DO I = 1,N
         Y0(I) = YHAT(I)
      END DO
      !check new step doesn't take us past B
      IF(ABS(X0 - A + HNEW).GT. ABS(B - A)) HNEW = B - X0
```

```
        H = HNEW

          !Save the data to file

          IF( ABS(Xo - A)/COUNT .GE. ABS(HMAX) )THEN

             WRITE(1,100) Xo, (Yo(I),I=1,N)

             COUNT = COUNT + 1

          END IF

        END IF

      END DO

100 FORMAT(999(2X,ES12.5))

      !Assign calculated values back to subroutine arguments

      X = Xo

      DO I = 1,N

         Y(I) = Yo(I)

      END DO

      !write final values to file

      WRITE(1,100) Xo, (Yo(I),I=1,N)

      !Add a blank line at end of data file before closing

      WRITE(1,*)

      CLOSE(1)

      RETURN

      END SUBROUTINE RKF45N


!*****************************************************************


      SUBROUTINE DERIV( N, X, Y, K )

        INTEGER N,I,N2

        DOUBLE PRECISION X, Y(N), K(N), DF

        EXTERNAL DF

        N2 = N/2
```

Appendix

```fortran
      DO I = 1,N2

        K(I) = Y(I+N2)

        K(I+N2) = DF( X, Y, I )

      END DO

      RETURN

    END SUBROUTINE DERIV

!*************************************************************

  END MODULE RKF45_MODULE

!END OF FILE ****************************************************
```

rkf45FFT_mod.f90

```fortran
!*************************************************************
!Module containing Runge-Kutta-Fehlberg algorithm to produce
!data for the FFT subroutine.
!*************************************************************

MODULE RKF45FFT_MODULE

  IMPLICIT NONE

  CONTAINS

!*************************************************************

    SUBROUTINE RKF45FFT( DF, X, Y, A, B, TOL, DATA, DELTA )

      !*********************************************************
      !RK4F2 uses a fourth/fifth order Runge-Kutta-Fehlberg    *
      !algorithm to solve second order ODEs.                   *
      !2nd order ODE -> pair of coupled 1st order ODEs         *
      !*********************************************************

      !Declare subroutine arguments
      DOUBLE PRECISION DF, X, Y(2), A, B, TOL, DELTA
      EXTERNAL DF

      !Declare variables for FFT sampling
```

```fortran
DOUBLE PRECISION, PARAMETER :: GOAL = 100

INTEGER,        PARAMETER :: NFFT = 512

COMPLEX*16 :: DATA(NFFT*2)

INTEGER IFFT

IFFT = 1

!Define the conservative factor for new step size

DOUBLE PRECISION, PARAMETER :: ALPHA = 9.D-1

!FRACMAX: Fraction of interval defining HMAX (B-A/FRACMAX)

!FRACMIN: Fraction of interval defining HMIN (B-A/FRACMIN)

DOUBLE PRECISION, PARAMETER :: FRACMAX = 1.D1

DOUBLE PRECISION, PARAMETER :: FRACMIN = 1.D5

!INC: Max tolerated increase factor of h

!DEC: Max tolerated decrease factor of h

DOUBLE PRECISION, PARAMETER :: INC = 4.D0

DOUBLE PRECISION, PARAMETER :: DEC = 1.D-1

! Am define the coefficients used to calculate x

DOUBLE PRECISION, PARAMETER :: A1 = 2.5D-1

DOUBLE PRECISION, PARAMETER :: A2 = 3.75D-1

DOUBLE PRECISION, PARAMETER :: A3 = 1.2D1/1.3D1

DOUBLE PRECISION, PARAMETER :: A4 = 1.D0

DOUBLE PRECISION, PARAMETER :: A5 = 5.D-1

!Bmn define the coefficients used to calculate y

DOUBLE PRECISION, PARAMETER :: B10 = 2.5D-1

DOUBLE PRECISION, PARAMETER :: B20 = 3.D0/3.2D1

DOUBLE PRECISION, PARAMETER :: B21 = 9.D0/3.2D1

DOUBLE PRECISION, PARAMETER :: B30 = 1.932D3/2.197D3

DOUBLE PRECISION, PARAMETER :: B31 = -7.2D3/2.197D3

DOUBLE PRECISION, PARAMETER :: B32 = 7.296D3/2.2197D3

DOUBLE PRECISION, PARAMETER :: B40 = 4.39D2/2.16D2
```

Appendix

DOUBLE PRECISION, PARAMETER :: B41 = -8.D0

DOUBLE PRECISION, PARAMETER :: B42 = 3.68D3/5.13D2

DOUBLE PRECISION, PARAMETER :: B43 = -8.45D2/4.104D3

DOUBLE PRECISION, PARAMETER :: B50 = -8.D0/2.7D1

DOUBLE PRECISION, PARAMETER :: B51 = 2.D0

DOUBLE PRECISION, PARAMETER :: B52 = -3.544D3/2.565D3

DOUBLE PRECISION, PARAMETER :: B53 = 1.859D3/4.104D3

DOUBLE PRECISION, PARAMETER :: B54 = -1.1D1/4.D1

!Cm are used to evaluate YHAT

DOUBLE PRECISION, PARAMETER :: C0 = 1.6D1/1.35D2

DOUBLE PRECISION, PARAMETER :: C2 = 6.656D3/1.2825D4

DOUBLE PRECISION, PARAMETER :: C3 = 2.8561D4/5.643D4

DOUBLE PRECISION, PARAMETER :: C4 = -9.D0/5.D1

DOUBLE PRECISION, PARAMETER :: C5 = 2.D0/5.5D1

!Dm are used to calculate the error

DOUBLE PRECISION, PARAMETER :: D0 = 1.D0/3.6D2

DOUBLE PRECISION, PARAMETER :: D2 = -1.28D2/4.275D3

DOUBLE PRECISION, PARAMETER :: D3 = -2.197D3/7.524D4

DOUBLE PRECISION, PARAMETER :: D4 = 1.D0/5.D1

DOUBLE PRECISION, PARAMETER :: D5 = 2.D0/5.5D1

!Declare local variables

DOUBLE PRECISION H, HMAX, HNEW, HMIN, X0, ERR, BIGERR

DOUBLE PRECISION, DIMENSION(2) :: Y0, YHAT, K0, K1

DOUBLE PRECISION, DIMENSION(2) :: K2, K3, K4, K5

INTEGER I

IF(A.GE.B)STOP 'User error: B must be .gt. A'

IF(TOL.LT.EPSILON(0.D0))THEN

  PRINT*, 'User error: Tol must be .gt.',EPSILON(0.D0)

  STOP

```
END IF

!Set maximum step size

HMAX = (B - A)/FRACMAX

!Set minimum step size

HMIN = (B - A)/FRACMIN

!Initialise the integration;

!initial Y values defined in calling program.

H = HMAX

X0 = A

DO I = 1,2
  Y0(I) = Y(I)
END DO

! save initial value to data array

DATA(IFFT) = Y(1)

!Perform the integration; exit after collecting required data

DO WHILE (IFFT.LT.NFFT)
  CALL DERIV( X0, Y0, K0 )
  X = X0 + A1*H
  DO I = 1,2
    Y(I) = Y0(I) + B10*H*K0(I)
  END DO
  CALL DERIV( X, Y, K1 )
  X = X0 + A2*H
  DO I = 1,2
    Y(I) = Y0(I) + H*(B20*K0(I) + B21*K1(I))
  END DO
  CALL DERIV( X, Y, K2 )
  X = X0 + A3*H
  DO I = 1,2
```

Appendix

```
    Y(I) = Yo(I) + H*(B30*K0(I) + B31*K1(I) &
        + B32*K2(I))
END DO
CALL DERIV( X, Y, K3 )
X = Xo + A4*H
DO I = 1,2
    Y(I) = Yo(I) + H*(B40*K0(I) + B41*K1(I) &
        + B42*K2(I) + B43*K3(I))
END DO
CALL DERIV( X, Y, K4 )
X = Xo + A5*H
DO I = 1,2
    Y(I) = Yo(I) + H*(B50*K0(I) + B51*K1(I) &
        + B52*K2(I) + B53*K3(I)  + B54*K4(I))
END DO
CALL DERIV( X, Y, K5 )
BIGERR = 0.D0
DO I = 1,2
    YHAT(I) = Yo(I) + H*(C0*K0(I) + C2*K2(I) &
        + C3*K3(I) + C4*K4(I)  + C5*K5(I))
    ERR = H * ABS(D0*K0(I) + D2*K2(I) + D3*K3(I) &
        + D4*K4(I) + D5*K5(I))
    IF(ERR .GT. BIGERR)BIGERR = ERR
END DO
HNEW = ALPHA * H * SQRT( SQRT( H*TOL/BIGERR ) )
!Check prediction of step size falls within
!acceptable limits and adjust if necessary
IF( HNEW .GT. INC * H ) HNEW = INC * H
IF( HNEW .LT. DEC * H ) HNEW = DEC * H
```

```
    IF( HNEW .GT. HMAX) HNEW = HMAX

    IF( HNEW .LT. HMIN)THEN

      PRINT *, H,'->', HNEW

      WRITE (1,100) X, YHAT(1), YHAT(2), YHAT(3), YHAT(4)

      STOP 'Step size too small; exiting integration'

    END IF

    !Check error vs. max_error = h*eps

    !if error too large repeat step using hnew

    !else accept propagation

    IF( BIGERR .GT. (H*TOL) )THEN

      H = HNEW

    ELSE

      X0 = X0 + H

      IF( ABS((X0-GOAL)/X0) .LT. 1.D-5)THEN

        IFFT = IFFT + 1

        DATA(IFFT) = YHAT(1)

        GOAL = GOAL + DELTA

      END IF

      IF(H .GT. GOAL-X0) H = GOAL-X0

      DO I = 1,2

        Y0(I) = YHAT(I)

      END DO

      H = HNEW

    END IF

  END DO

100 FORMAT(5(2X,F10.6))

    !Assign calculated values back to subroutine arguments

    X = X0

    DO I = 1,2
```

Appendix

```
      Y(I) = Yo(I)

   END DO

   !Add a blank line at end of data file before closing

   WRITE(1,*)

   CLOSE(1)

   RETURN

   END SUBROUTINE RKF45FFT

!***********************************************************

   SUBROUTINE DERIV( X, Y, K )

      DOUBLE PRECISION X, Y(2), K(2), DF

      EXTERNAL DF

      K(1) = Y(2)

      K(2) = DF( X, Y )

      RETURN

   END SUBROUTINE DERIV

!***********************************************************

   END MODULE RKF45FFT_MODULE

!END OF FILE************************************************
```

Chapter 12 – High Performance Computing

matrixAdd.f90

!!$ Program to test the performance of array addition when

!!$ row and column indices are swapped.

```fortran
PROGRAM MATRIXADD

 USE OMP_LIB

 IMPLICIT NONE

 INTEGER,    PARAMETER  :: N = 1000

 DOUBLE PRECISION, ALLOCATABLE :: A(:,:),B(:,:),C(:,:)

 DOUBLE PRECISION T1,T2,T3

 INTEGER I,J

 ALLOCATE( A(N,N), B(N,N), C(N,N) )

 A = 1.D0

 B = 2.D0

 T1 = OMP_GET_WTIME()

 DO J = 1,N

   DO I = 1,N

     C(I,J) = A(I,J) + B(I,J)

   END DO

 END DO

 T2 = OMP_GET_WTIME()

 DO I = 1,N

   DO J = 1,N

     C(I,J) = A(I,J) + B(I,J)

   END DO

 END DO

 T3 = OMP_GET_WTIME()

 PRINT *, 'Time for JI loop:',T2-T1
```

```
    PRINT *, 'Time for IJ loop:',T3-T2

    DEALLOCATE( A, B, C )

END PROGRAM MATRIXADD

!END OF FILE*********************************************
```

matrixMul1.f90

```
!***********************************************************
! Program to test the performance of the naive approach to
! matrix multiplication.
!***********************************************************

PROGRAM MATRIXMUL1

    USE OMP_LIB

    IMPLICIT NONE

    DOUBLE PRECISION, ALLOCATABLE :: A(:,:), B(:,:), C(:,:)

    DOUBLE PRECISION T1, T2

    INTEGER N, I, J, K

    INTEGER, PARAMETER :: N_MAX = 200

    OPEN(UNIT=1, FILE='matrix1_times.txt', ACTION='WRITE')

100 FORMAT(I4,2X,ES12.6)

    DO N = N_MAX,N_MAX

      ALLOCATE ( A(N,N), B(N,N), C(N,N) )

      DO J = 1,N

        DO I = 1,N

          A(I,J) = I+J

          B(I,J) = J

          C(I,J) = 0.0

        END DO

      END DO
```

```fortran
      T1 = OMP_GET_WTIME()


      CALL MATMUL(A,B,C,N)

      T2 = OMP_GET_WTIME()

      WRITE(1,100), N, T2-T1

      PRINT *, T2-T1

      DEALLOCATE( A, B, C )

      PRINT *, '% done:',N*100/N_MAX
    END DO
    CLOSE(1)
END PROGRAM MATRIXMUL1
!*******************************************************
SUBROUTINE MATMUL(A,B,C,N)
  INTEGER N, I, J, K
  DOUBLE PRECISION A(N,N), B(N,N), C(N,N)
  DO I = 1,N
    DO J = 1,N
      DO K = 1,N
        C(I,J) = C(I,J) + A(I,K)*B(K,J)
      END DO
    END DO
  END DO
  RETURN
END SUBROUTINE MATMUL
!END OF FILE *****************************************
```

Appendix

matrixMul2.f90

!**

! Program to test the performance of the naive approach to

! matrix multiplication.

!**

```fortran
PROGRAM MATRIXMUL2
  USE OMP_LIB
  IMPLICIT NONE
  DOUBLE PRECISION, ALLOCATABLE :: A(:,:), B(:,:), C(:,:)
  DOUBLE PRECISION T1, T2
  INTEGER I, J, K, NB, M
  INTEGER IMIN, IMAX, JMIN ,JMAX, KMIN, KMAX
  INTEGER, PARAMETER :: N = 2**9
  OPEN(UNIT=1, FILE='matrix2_times.txt', ACTION='WRITE')
  100 FORMAT(I4,2X,F10.6)
  ALLOCATE ( A(N,N), B(N,N), C(N,N) )
  DO J = 1,N
    DO I = 1,N
      A(I,J) = I+J
      B(I,J) = J
      C(I,J) = 0.0
    END DO
  END DO
  M = 2**3
  NB = N/M
  T1 = OMP_GET_WTIME()
  DO K = 1,M
    KMIN = (K-1)*NB + 1
    KMAX = K*NB
```

```fortran
     DO I = 1,M
       IMIN = (I-1)*NB + 1
       IMAX = I*NB
       DO J = 1,M
         JMIN = (J-1)*NB + 1
         JMAX = J*NB
         CALL MATMUL( A(IMIN:IMAX, KMIN:KMAX), &
           B(KMIN:KMAX, JMIN:JMAX), &
           C(IMIN:IMAX, JMIN:JMAX), NB )
       END DO
     END DO
   END DO
   T2 = OMP_GET_WTIME()
   CALL DGEMM('N', 'N', N, N, N, 1.D0, A, N, B, N, 0.D0, C, N)
   T3 = OMP_GET_WTIME()
   WRITE(1,100), NB, T2-T1
   PRINT *,'Block time:', T2-T1
   PRINT *,'LAPACK time:', T3-T2
   DEALLOCATE( A, B, C )
   CLOSE(1)
 END PROGRAM MATRIXMUL2
!********************************************************
 SUBROUTINE MATMUL(AS,BS,CS,N)
   INTEGER N, I, J, K
   DOUBLE PRECISION AS(N,N), BS(N,N), CS(N,N)
   DO I = 1,N
     DO J = 1,N
       DO K = 1,N
         CS(I,J) = CS(I,J) + AS(I,K)*BS(K,J)
```

```
      END DO

    END DO

  END DO

  RETURN

END SUBROUTINE MATMUL

!END OF FILE *****************************************
```

loopRoll.f90

```
!!$ Program to test the performance loop unrolling
PROGRAM LOOPROLL

  USE OMP_LIB

  IMPLICIT NONE

  INTEGER,      PARAMETER   :: N = 250, MMAX = 1000

  DOUBLE PRECISION, DIMENSION (N,N) :: A,B,C

  DOUBLE PRECISION T1,T2,T3,T4,TA,TB,TC

  INTEGER I,J,M

  A = 1.D0

  B = 2.D0

  M = 1

  TA = 0.D0

  TB = 0.D0

  TC = 0.D0

  DO WHILE (M .LE. MMAX)

    T1 = OMP_GET_WTIME()

    DO J = 1,N

      DO I = 1,N

        C(I,J) = A(I,J) + B(I,J)

      END DO

    END DO
```

```fortran
      T2 = OMP_GET_WTIME()
      DO J = 1,N
        DO I = 1,N,2
          C(I,J)   = A(I,J)   + B(I,J)
          C(I+1,J) = A(I+1,J) + B(I+1,J)
        END DO
      END DO
      T3 = OMP_GET_WTIME()
      DO J = 1,N
        DO I = 1,N,4
          C(I,J)   = A(I,J)   + B(I,J)
          C(I+1,J) = A(I+1,J) + B(I+1,J)
          C(I+2,J) = A(I+2,J) + B(I+2,J)
          C(I+3,J) = A(I+3,J) + B(I+3,J)
        END DO
      END DO
      T4 = OMP_GET_WTIME()
      TA = TA + T2 - T1
      TB = TB + T3 - T2
      TC = TC + T4 - T3
      M = M + 1
      END DO
      PRINT *, 'Time loop 1:',TA
      PRINT *, 'Time loop 2:',TB
      PRINT *, 'Time loop 3:',TC
      END PROGRAM LOOPROLL
      !END OF FILE*****************************************************
```

OMP_HelloWorld.f90

```fortran
PROGRAM HELLO

  USE OMP_LIB

  IMPLICIT NONE

  INTEGER NTHREADS, TID

  !Fork a team of threads giving them their own copies of variables

  !$OMP PARALLEL PRIVATE(NTHREADS, TID)

  !Obtain thread number

  TID = OMP_GET_THREAD_NUM()

  PRINT *, 'Hello World from thread = ', TID

  !Only master thread does this

  IF (TID .EQ. 0) THEN

    NTHREADS = OMP_GET_NUM_THREADS()

    PRINT *, 'Number of threads = ', NTHREADS

  END IF

  !All threads join master thread and disband

  !$OMP END PARALLEL

END PROGRAM HELLO

!END OF FILE*********************************************
```

omp_param.f90

```fortran
PROGRAM OMPPARAM

  USE OMP_LIB

  IMPLICIT NONE

  PRINT *, 'Max threads', OMP_GET_MAX_THREADS()

  PRINT *, 'Num of cores', OMP_GET_NUM_PROCS()

END PROGRAM OMPPARAM

!END OF FILE*********************************************
```

vectorSum.f90

```fortran
PROGRAM VECSUM
  USE OMP_LIB
  IMPLICIT NONE
  DOUBLE PRECISION, ALLOCATABLE :: A(:)
  INTEGER, PARAMETER :: N = 10000000, MMAX = 100
  INTEGER I, J, TID, M, NUM
  DOUBLE PRECISION T1,T2,T3,T4,TS,TP,SUM
  CALL OMP_SET_NUM_THREADS(4)
  ALLOCATE( A(N) )
  DO I = 1,N
    A(I) = 1.D0
  END DO
  PRINT *, 'SINGLE CORE...'
  M = 1
  TS = 0.D0
  TP = 0.D0
  DO WHILE (M .LE. MMAX)
    SUM = 0.D0
    T1 = OMP_GET_WTIME()
    DO I = 1,N
      SUM = SUM + A(I)
    END DO
    T2 = OMP_GET_WTIME()
    TS = TS + (T2 - T1)
    M = M + 1
  END DO
  SUM = 0.D0
```

Appendix

```fortran
      M = 1
      PRINT *, 'PARALLEL CORES...'
      DO WHILE (M .LE. MMAX)
        T3 = OMP_GET_WTIME()
        !$OMP PARALLEL SHARED(A,SUM) PRIVATE(I)
        !$OMP DO REDUCTION(+:SUM)
        DO I = 1,N
          SUM = SUM + A(I)
        END DO
        !$OMP END DO
        !$OMP END PARALLEL
        T4 = OMP_GET_WTIME()
        TP = TP + T4 - T3
        M = M + 1
      END DO
      PRINT *, 'Serial time =', TS/DBLE(MMAX)
      PRINT *, 'Parallel time =', TP/DBLE(MMAX)
      DEALLOCATE( A )
      END PROGRAM VECSUM
      !END OF FILE*****************************************
```